Smart Electricity Distribution Networks

Smart Electricity Distribution Networks

Chengshan Wang, Jianzhong Wu,
Janaka Ekanayake and Nick Jenkins

CRC Press
Taylor & Francis Group
Boca Raton London New York

CRC Press is an imprint of the
Taylor & Francis Group, an **informa** business

MATLAB® and Simulink® are trademarks of the MathWorks, Inc. and are used with permission. The MathWorks does not warrant the accuracy of the text or exercises in this book. This book's use or discussion of MATLAB® and Simulink® software or related products does not constitute endorsement or sponsorship by the MathWorks of a particular pedagogical approach or particular use of the MATLAB® and Simulink® software.

Taylor & Francis Group
6000 Broken Sound Parkway NW, Suite 300
Taylor & Francis Group
Boca Raton, FL 33487-2742

First issued in paperback 2020

© 2017 by Taylor & Francis Group, LLC
CRC Press is an imprint of Taylor & Francis Group, an Informa business

No claim to original U.S. Government works

ISBN 13: 978-0-367-57387-4 (pbk)
ISBN 13: 978-1-4822-3055-0 (hbk)

Library of Congress Cataloging-in-Publication Data

Names: Wang, Chengshan, 1962- author. | Wu, Jianzhong, 1976- author. | Ekanayake, Janaka, 1964- author. | Jenkins, Nicholas, 1954- author.
Title: Smart electricity distribution networks / Chengshan Wang, Jianzhong Wu, Janaka Ekanayake, Nick Jenkins.
Description: Boca Raton : Taylor & Francis, a CRC title, part of the Taylor & Francis imprint, a member of the Taylor & Francis Group, the academic division of T&F Informa, plc, [2017] | Includes bibliographical references and index.
Identifiers: LCCN 2016049431 | ISBN 9781482230550 (hardback : alk. paper) | ISBN 9781482230567 (ebook)
Subjects: LCSH: Smart power grids.
Classification: LCC TK3105 .S48575 2017 | DDC 621.319/2–dc23
LC record available at https://lccn.loc.gov/2016049431

Visit the Taylor & Francis Web site at
http://www.taylorandfrancis.com

and the CRC Press Web site at
http://www.crcpress.com

CONTENTS

9   **Energy Management and Optimal Planning of Microgrids**   **293**

PREFACE

Smart grid involves modernizing electric power networks and changing the way they are planned and operated. It empowers energy consumers, provides new services and supports the transition to a sustainable low-carbon economy. Although the driving forces and the objectives of smart grid development are different in different countries, smart distribution networks are always one of the key research and development topics.

The use of distributed energy resources (DERs), e.g. distributed generation, flexible loads and energy storage, is increasing rapidly in many countries and this can lead to operational difficulties in distribution networks. The changes that are already occurring in distribution networks include a dramatic increase in intermittent distributed generation while new home energy management and electric vehicle charging systems are introducing new patterns of load. Power flows in the distribution network are becoming more volatile and unpredictable as new types of generation and loads are connected, and customers gain greater understanding and control of their loads. In response to these changes, sophisticated measurement, control and ICT systems are being deployed throughout the network, and new incentives and market arrangements are emerging.

These changes to the distribution system will accelerate over the coming decades and the behaviour of distribution networks will become increasingly uncertain as new smart grid and smart energy interventions are introduced. Hence, there is a growing need for academia, industry and policymakers to move towards new planning and operation approaches to electricity distribution networks considering emerging technologies and uncertainties.

The aim of this book is to provide a fundamental discussion of smart distribution networks and the new technologies associated with them. The topics that are covered include an introduction to the new technologies in electricity distribution, which include various kinds of generation connected at the distribution level. Different concepts to manage a high penetration of DERs are discussed, including demand-side integration, microgrids, CELLs and virtual power plants. From the perspective of distribution network development, information and communication infrastructure and new devices are addressed. Methods to analyse smart distribution systems are then described, including both steady state and transient analysis. The operation and planning of smart distribution networks are reviewed and DC distribution networks discussed.

This book will be valuable to all those who want to understand the key enabling technologies and performance of smart electricity distribution networks. It will allow readers to engage with the immediate development of electricity distribution networks and take part in the wider debate over the future smart grid.

MATLAB® and Simulink® are registered trademarks of The MathWorks, Inc. For product information, please contact:

The MathWorks, Inc.
3 Apple Hill Drive
Natick, MA, 01760-2098 USA
Tel: 508-647-7000
Fax: 508-647-7001
E-mail: info@mathworks.com
Web: www.mathworks.com

ACKNOWLEDGEMENTS

The authors acknowledge the support of many individuals and organizations, without whom this book would not have been possible.

The underlying research was funded by the UK–China EPSRC/NFSC joint project 'Integrated Operation and Planning for Smart Electric Distribution Networks (OPEN)', and the National 973 Program of China 'Research on the Key Issues of Distributed Generation Systems'.

Specific sections of the book arose from several other collaborative projects and case studies, which include the European Commission Horizon 2020 project 'Peer to Peer Smart Energy Distribution Networks', EPSRC projects 'MISTRAL: Multi-Scale Infrastructure Systems Analytics' and 'Increasing the Observability of Electrical Distribution Systems using Smart Meters', the NIC project ANGLE-DC, the National 863 Program of China 'Research and Development of Self-Healing Control Technologies of Smart Distribution System', the NFSC project 'Protection and Control of Active Electrical Distribution Network' and the NRC (Sri Lanka) project 'DC networks for Energy Efficiency and Renewable Additions'.

We have learnt from collaborations and conversations with a number of industrial and public sector partners, including the State Grid of China, China Southern Power Grid, National Grid of the United Kingdom, Scottish Power Energy Networks, Toshiba Europe TRL, Electric Corby and others.

We acknowledge contributions from colleagues and individuals. We thank Dr. Xialin Li, Dr. Hao Yu, Dr. Chongbo Sun, Dr. Bingqi Jiao, Yixin Liu at Tianjin University and Dr. Meng Cheng at Cardiff University, who helped in numerous ways.

AUTHORS

Professor Chengshan Wang received his BSc, MSc and PhD in electrical engineering in 1983, 1985 and 1991, respectively, from Tianjin University, China. He has been with Tianjin University from 1985 and has been a professor in the university since 1996. He is the dean of the School of Electrical Engineering and Automation, Tianjin University. He is Chang-Jiang (Cheung Kong) Scholar Professor at Tianjin University and the director of the Key Laboratory of Smart Grid of the Ministry of Education in China. Professor Wang was with Cornell University as a visiting scientist from 1994 to 1996 and with Carnegie Mellon University as a visiting professor from 2001 to 2002. He received three Chinese National Awards of Science and Technology in 2004, 2010 and 2012. He received seven awards from the Education Ministry and some provincial governments in China for his excellent research work. He is also a recipient of the Fok Ying Tung Fund, the Science and Technology Achievement Award of the Ho Leung Ho Lee Foundation and the National Science Fund for Distinguished Young Scholars. Professor Wang's research is generally concerned with the operation, management, and planning of distribution systems and sustainable electric energy systems. In particular, his research interests include the distributed generation system and microgrids. He was the principal investigator of the Chinese National Basic Research Program (973) 'Research on the Key Issues of Distributed Generation Systems' from 2009 to 2013. Professor Wang is an active volunteer with IEEE Power and Energy Society (PES) and has served as the technical programme chair of the 2012 IEEE Innovative Smart Grid Technologies Asia Conference, editor for *IEEE Transactions on Sustainable Energy*, guest editor for *IEEE Transactions on Smart Grid* (Special Issue on 'Smart Grid Technologies and Development in China') and guest editor for *Applied Energy* (Special Issue on 'Smart Grids'). Professor Wang is the lead author of Chapters 1, 3, 6, 7, 8, 9 and 10 and has edited the whole book.

Professor Jianzhong Wu is a professor of multi-vector energy systems at Cardiff University. He joined Cardiff University as a lecturer in June 2008 and was promoted to senior lecturer (2013), reader (2014) and professor (2015). From 2006 to 2008, he was a research fellow at the University of Manchester. Professor Wu researches smart grid and energy infrastructure. He has a track record of undertaking a number of large research projects in smart grids and energy infrastructure. He has been principal investigator or co-investigator of more than 30 research projects funded by the European Commission, the Research Council of the United Kingdom and the industry. His professional involvements include subject editor of *Applied Energy* (since 2015); guest editor-in-chief of *Special Issues on Integrated Energy Systems* (2015) and on *Synergies between Energy Networks for Applied Energy* (2016); director of Applied Energy UNiLAB on Synergies between Energy Networks (since 2016); vice-chair of the organizing committee of the 2012 IEEE PES Innovative Smart Grid Technologies (ISGT Asia), 2012; and a member of the organizing committees of Applied Energy Symposium and Forum REM2016, CUE2015, and the Fifth, Sixth, Seventh, and Eighth International Conferences on Applied Energy (ICAE). He has published widely in these areas. He received his PhD from Tianjin University, China, in 2004. Professor Wu is the lead author of Chapter 3, contributed to Chapters 1, 2, 5, 6, 7, 8, 9 and 10 and edited the whole book.

Professor Janaka Ekanayake has been affiliated with the Department of Electrical and Electronic Engineering, University of Peradeniya, Sri Lanka, as a professor since April 2013. He is also a visiting professor at the Institute of Energy at Cardiff University. Prior to that, until end of October 2012, he was a senior research fellow/reader at the Institute of Energy, Cardiff University, United Kingdom. He is a fellow of IET (United Kingdom) and IESL (Sri Lanka) and a senior member of the IEEE. He is also recognized as an IEEE PES distinguished lecturer. His main research interests include power electronic applications for power system, renewable energy generation and its integration and smart grids. He has published more than 50 papers in refereed journals and has co-authored 5 books. The key books to which he contributed are *Electric Power Systems* (2012), Wiley; *Smart Grid: Technology and Applications* (2012), Wiley; *Distributed Generation* (2010), Institution of Engineering and Technology; and *Wind Energy Generation: Modelling and Control* (2009), Wiley. He was a Tyndall Centre Research Fellow (2002 and 2003), Commonwealth Fellow (2001–2002), and a Royal Society of UK Research Fellow (1997) at the University of Manchester Institute of Science and Technology (UMIST), United Kingdom. He is a member of the editorial board of *IEEE Transactions on Energy Conversion* (2007 to date), *IET Journal of Renewable Energy* (2015 to date) and *Journal of Wind Energy* (2013 to date). He was also the organising vice chairperson of the First IEEE PES Conference of Innovative Smart Grid Technologies (2012). Professor Ekanayake is the lead author of Chapters 2, 4, 5 and 11, and he contributed to Chapters 8 and 10.

Professor Nick Jenkins joined the University of Manchester (UMIST) in 1992 after 14 years in industry. He was appointed professor in 1998 and moved to Cardiff University in 2008. He has developed teaching and research activities in both electrical power engineering and renewable energy. He is a fellow of the IET, IEEE, and the Royal Academy of Engineering and is a distinguished member of CIGRE. He served on the DECC/OFGEM Smart Grids Forum and the OFGEM Low Carbon Network Fund Panel. From 2008 to 2011, he was Shimizu Visiting Professor at Stanford University. Professor Nick Jenkins has contributed to more than 15 books on renewable energy generation and power systems. He contributed to Chapters 1, 2, 5, 10 and 11.

CONTRIBUTING AUTHORS

Dr. Prabath Binduhewa is a senior lecturer attached to the Department of Electrical and Electronic Engineering, University of Peradeniya Sri Lanka (2010 to date). In 2010, he earned his PhD in electrical and electronic engineering from the University of Manchester, United Kingdom. His PhD thesis was on developing a microsource interface for a microgrid with energy storage. His research interests are in power electronic applications in renewable energy, energy storage systems, microgrids and power systems. He has co-authored more than 25 publications. He is member of the IEEE. Dr. Binduhewa contributed to Chapter 11.

Dr. Xiaopeng Fu received his BSc and PhD in electrical engineering in 2011 and 2016 from Tianjin University, China. He is currently a research associate at Tianjin University. His research interest is simulation analysis of power system transients. Dr. Fu contributed to Chapter 7 of the book.

Dr. Bin Li earned his BSc, MSc and PhD in electrical engineering from Tianjin University in 1999, 2002 and 2005, respectively. He then joined Tianjin University in 2006 as a professor. In the same year, he was academic visitor at the University of Manchester, United Kingdom. From 2008 to 2009, he worked in the design and application of protection relays and phasor measurement unit as a BOND engineer, in AREVA Company, United Kingdom. He is a winner of the National Outstanding Youth Fund of China (2014) and the New Century Talents by the Ministry of Education of China (2011). Professor Li's main research focuses on the protection and control of smart grids. Currently, he is an investigator in several ongoing research projects in this area supported by the National Natural Science Foundation of China and overseas and domestic industrial companies. Professor Li has published five books as co-author and more than 100 papers. He is the key author of Chapter 8.

Dr. Peng Li is an associate professor at Tianjin University. He received his BSc and PhD in electrical engineering in 2004 and 2010 from Tianjin University. Dr. Peng Li's research interests include distributed generation systems and microgrids, simulation analysis of power system transients and operation and planning of smart distribution systems. Dr. Li is the key author of Chapters 1, 6 and 7.

Dr. Hong Liu is an associate professor at Tianjin University. He received his BSc, MSc and PhD in electrical engineering in 2002, 2005 and 2009 from Tianjin University, China. Dr. Liu's research interests include optimal planning and reliability assessment of power distribution systems and integrated energy systems, grid integration of distributed generation and electric vehicle charging stations. Dr. Liu is the key author of Chapter 10.

Dr. Chao Long received his BSc from Wuhan University, China, in 2008 and his PhD from Glasgow Caledonian University, United Kingdom, in 2014. He was a postdoc research associate at the University of Manchester, United Kingdom, from August 2013 to May 2015 and is currently a postdoc research associate at Cardiff University, United Kingdom. Dr. Long contributed to Chapters 3, 5 and 10.

Dr. Lilantha Samaranayake is a lecturer in the Department of Electrical and Electronic Engineering, Faculty of Engineering, University of Peradeniya, Sri Lanka (2006 to date). He served as a senior scientist in the National Nanotechnology Initiative of Sri Lanka (2009 to 2011) and is a research fellow in Control of Electric Machines in the Advanced Vehicle Engineering Centre of Cranfield University, United Kingdom (2014 to date). His research involves electric vehicle powertrains, machine control, battery supercapacitor hybrid systems, DC grids and multi-terminal HVDC systems. He has co-authored 40 publications and holds two U.S. patents. He is a senior member of the IEEE. Dr. Samaranayake contributed to Chapter 4.

Dr. Yue Zhou received his BSc and PhD in electrical engineering in 2011 and 2016 from Tianjin University, China. He is currently a postdoc at Cardiff University. Dr. Zhou's research interests include home energy management, demand response and optimization methods. Dr. Zhou contributed to the editing of all chapters and the questions/solutions in this book.

CHAPTER 1

Introduction

1.1 Smart Distribution Networks

As an important form of energy, electricity offers the advantages of being clean, highly efficient and convenient for users. The power system that links the generation, transmission, distribution and consumption of electricity is one of the most complex man-made systems constructed to date. With the increasing demand for energy throughout the world and the associated environmental problems in recent years, conventional centralized power systems are facing significant challenges [1,2]. The development of a highly efficient and environmentally friendly smart grid has become an important objective worldwide.

Smart distribution networks will play an important role in future smart energy systems in providing a link between the transmission grid and the consumers [3]. A smart distribution network is the integration of advanced distribution automation, distributed generation and microgrid technologies. Advanced information, communication and computation technologies are essential in a smart distribution network to support its planning, operation and control. As a result, the smart distribution network becomes a complex cyber-physical system that connects energy and information networks [4]. Using smart distribution terminal units, a smart distribution network is able to ensure optimized operation under normal operating conditions and self-heal when faults occur. It can provide safe, reliable, high-quality, economic and environmentally friendly electrical power [5].

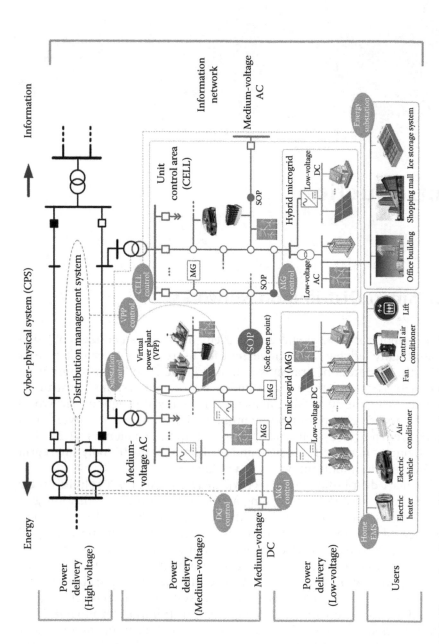

FIGURE 1.1 Illustrative diagram of a smart distribution network.

A typical configuration of a smart distribution network is shown in Figure 1.1. A distribution network can be divided into high, medium and low voltage levels. For medium and high voltages, the network topology has evolved from a simple radial arrangement into meshed or interconnected networks, in order to ensure high reliability, operational economy and equipment utilization. In addition to being connected to the bulk power system, power for a smart distribution network is also supplied from distributed generators (DGs) located at the medium and low voltage levels [6]. In this way, the smart distribution network is able to take full advantage of renewable energy resources. Moreover, future smart distribution networks will consist of both AC and DC networks, which will better serve a large number of DC loads, such as computers and electric vehicles (EVs) [7].

1.2 New Characteristics of a Smart Distribution Network

1.2.1 Distributed Energy Systems

A major challenge faced by smart distribution networks is the diversity in the types of DGs [8], which include photovoltaic units, wind power systems, fuel cells and micro-turbines, each DG utilizing different energy sources and showing different characteristics. Due to random fluctuations in some of the resources, for example wind and solar, energy storage systems (ESSs) are added to ensure the instantaneous and short-term power balance of the entire energy system [9,10]. These ESSs can be divided into electrochemical types, such as lead-acid, lithium and sodium–sulphur batteries; mechanical types, such as flywheels and compressed air ESSs; and electrical types, such as supercapacitors and superconducting magnetic ESSs. Most of these distributed energy resources are connected to the distribution network through power electronic converters, thus providing increased controllability (but reducing the inertia of the power system).

Besides these inputs from distributed energy resources, a large number of new controllable loads, such as EVs and smart home heaters, are emerging. Using the demand–response technology, these controllable loads serve as equivalent (negative) energy sources and participate in the optimal operation of the entire power system [11]. Distributed energy resources are described in detail in Chapter 2.

1.2.2 Multi-Layer Autonomous Operation Areas

In a smart distribution network, the components can be organized to form autonomous operating areas, in order to manage the DGs and serve the users' energy demands more effectively. These autonomous areas, such as microgrids [12], unit control areas (CELLs) [13] and virtual power plants (VPPs) [14], have different scales and operating objectives. A microgrid is a small, low-voltage power system unit that consists of DGs, ESSs, loads and monitoring and control systems; a CELL is an extension of a microgrid that covers a larger physical area at a higher voltage level; and a VPP controls the electricity demand and generation of a large area and by effective management of the controllable loads and DGs replicates the performance of a conventional power plant. Under normal conditions, these autonomous operating areas satisfy the internal load demand and are operated optimally. In emergency situations, they support each other and maintain power supplies to critical loads. An autonomous distribution area can be dispatched as a whole to facilitate optimized operation at the system level [15,16]. In Chapter 3, the basic concept and composition of microgrids, CELLs and VPPs are described. In Chapters 8 and 9, the operation, protection and control of a microgrid are discussed.

1.2.3 Information and Communication Systems

A powerful information and communication system is fundamental to a smart distribution network to manage tasks such as state awareness, information collection and broadcasting of commands. Compared with conventional distribution networks, the information and communication system of a smart distribution network has evolved considerably into a sophisticated assembly of advanced measurements, two-way, high-speed communication and big data management that stores and analyses information about power distribution and consumption. The results provide data fundamental for the planning and design, optimal operation, simulation and analysis of a smart distribution network. The information and communication technologies in smart distribution networks are discussed in detail in Chapter 4.

1.2.4 Novel Power Electronic Devices

Power electronic converters play an essential role in a smart distribution network and bring increased controllability of active and reactive power flows. The main function of power converters in smart distribution networks is to provide interfaces for DGs and ESSs. Depending on the characteristics and role of a DG or ESS, specific control strategies are designed for their associated power converters, such as constant active/reactive power control, constant voltage/frequency control and droop control. It is usual that these strategies are arranged to perform smooth switching between different operating conditions. In addition to providing interfaces for different types of DGs and ESSs, power converters can also be used as devices that replace traditional transformers and switches and provide services such as voltage regulation, power flow control, reactive power compensation and harmonic control. A soft open point (SOP) in a medium- or low-voltage network, which is often based on back-to-back power converters, is a typical example. It is used to replace a conventional switch connecting two feeders and thereby provides a way to optimize the operating condition of the distribution network by controlling the power flow between the two circuits [17]. In summary, the research and development of power electronic devices has become a critical issue in the development of smart distribution networks. Recent developments in novel power electronic devices are described in Chapter 5, and the use of direct current in distribution networks is introduced in Chapter 11.

1.3 Simulation of a Smart Distribution Network

Simulation is a fundamental tool for smart distribution networks. It plays an important role in the operational optimization, protection, control and R&D of novel distribution equipment. Depending on the time scale, simulations of a distribution network can be divided into steady state and transient studies. Steady-state simulation is based on power flow calculations and provides support to more advanced functions including short-circuit and reliability analysis, operational optimization, and self-healing control. To continuously describe changing processes, such as the state of charge (SOC) of batteries and the fluctuations of the wind and solar resource, steady-state simulations must have the ability to represent time series over multiple time scales including short term, daily and annual. In Chapter 6, the steady-state simulation of smart distribution networks is discussed.

Transient simulation algorithms need to be accurate and stable for non-linear problems in order to handle the properties of smart distribution network models such as the highly coupled dynamics of different energy sources and equipment. The simulation tools should provide access for flexible user-defined models to keep up with the rapid development of novel devices. The emergence of a large number of power electronic devices also presents challenges for transient simulations. Some special methods are required to resolve the issues of simulation accuracy and numerical stability that may be affected by the fast switching of power electronic devices. These issues are discussed in Chapter 7.

1.4 Operational Optimization of a Smart Distribution Network

Compared to conventional distribution networks, the use of actively controlled equipment is more common in smart distribution networks, providing methods to optimize network operation. Generally speaking, the operational optimization of smart distribution networks includes both the optimal dispatch of autonomous areas and the optimal control of various controllable devices. The optimization objectives include minimizing active/reactive power losses, energy losses, voltage excursions, and environmental costs. In some cases, these objectives may be combined together, forming a multi-objective optimization problem. The items that require optimization include the output of DGs, power drawn by controllable loads, status of switches, transformer tap positions, the operation of capacitor switching groups and SOP terminal power injections. The constraints include conventional limits such as system capacity, power flows, voltages and branch currents, as well as some new aspects such as DG output and battery states of charge.

Mathematically, the optimization of a smart distribution network is usually a large-scale, non-linear optimization problem. Based on different decision variables, this type of problem can be divided into two categories: continuous and discrete. The former solves continuous problems, such as DG output power optimization, and usually uses mathematical optimization algorithms; the latter solves discrete problems, such as the optimization of tie switch status, which are more suitable for artificial intelligence optimization techniques. For some complex, mixed-integer, strongly non-linear optimization problems, such as a network reconfiguration problem with SOPs, these optimization algorithms can be combined together to fully utilize the advantages of each type. The operational optimization of smart distribution networks is discussed in detail in Chapter 9.

1.5 Planning and Design of a Smart Distribution Network

Because of the need to integrate various DGs and diverse loads, the planning and design of smart distribution networks has shifted from the traditional simple network–load matching approach to a comprehensive source–network–load coordination problem. Both the complexity and difficulty of information management and decision-making have increased significantly across the whole planning and design process. In addition to traditional load forecasting, new factors, such as the capacities and spatial distributions of controllable loads and DGs, must be considered. In substation and network planning, novel devices such as DGs, ESSs, and SOPs play an important role in load shifting and balancing local loads, which may reduce or even reverse the power flow from the transmission grid. Additional tasks to be considered in the planning and design of smart distribution networks are to resolve issues such as the sizing and locating of DGs and ESSs, the design of distribution automation systems and the planning of communication systems. In addition to the traditional objectives of safety, reliability and economy, the planning of a smart distribution network must also consider new objectives including maximizing energy utilization efficiency, maximizing renewable energy utilization, minimizing environmental pollution and maximizing social benefits. These properties make the optimal planning of smart distribution networks a complex problem with multiple objectives and constraints. The relevant issues are introduced in Chapter 10.

☐ Questions

1. Explain the main functions of ESSs in a smart distribution network.

2. List the typical applications of power electronic converters in a smart distribution network.

3. Explain what requires optimization in the operation of a smart distribution network.

☐ References

1. A. Ipakchi and F. Albuyeh, Grid of the future, *IEEE Power and Energy Magazine*, 7(2), 52–62, March 2009.
2. H. Farhangi, The path of the smart grid, *IEEE Power and Energy Magazine*, 8(1), 18–28, January 2010.
3. R. F. Arritt and R. C. Dugan, Distribution system analysis and the future smart grid, *IEEE Transactions on Power Industry Applications*, 47(6), 2343–2350, November 2011.

4. S. K. Khaitan and J. D. Mccalley, Cyber physical system approach for design of power grids: A survey, *Proceedings of IEEE Power and Energy Society General Meeting*, July 2013, Vancouver, BC, Canada, pp. 1–5.

5. K. Moslehi and R. Kumar, A reliability perspective of the smart grid, *IEEE Transactions on Smart Grid*, 1(1), 57–64, June 2010.

6. J. A. P. Lopes, N. Hatziargyriou, J. Mutale, P. Djapic and N. Jenkins, Integrating distributed generation into electric power systems: A review of drivers, challenges and opportunities, *Electric Power Systems Research*, 77(9), 1189–1203, July 2007.

7. T. Dragicevic, J. C. Vasquez, J. M. Guerrero and D. Skrlec, Advanced LVDC electrical power architectures and microgrids: A step toward a new generation of power distribution networks, *IEEE Electrification Magazine*, 2(2), 54–65, March 2014.

8. J. Driesen and F. Katiraei, Design for distributed energy resources, *IEEE Power and Energy Magazine*, 6(3), 30–40, May 2008.

9. J. P. Barton and D. G. Infield, Energy storage and its use with intermittent renewable energy, *IEEE Transactions on Energy Conversion*, 19(2), 441–448, June 2004.

10. A. A. A. Radwan and Y. A. I. Mohamed, Assessment and mitigation of interaction dynamics in hybrid AC/DC distribution generation systems, *IEEE Transactions on Smart Grid*, 3(3), 1382–1393, September 2012.

11. P. Palensky and D. Dietrich, Demand side management: Demand response, intelligent energy systems, and smart loads, *IEEE Transactions on Industrial Informatics*, 7(3), 381–388, August 2011.

12. N. Hatziargyriou, H. Asano, R. Iravani and C. Marnay, Microgrids, *IEEE Power and Energy Society*, 5(4), 78–94, July 2007.

13. P. Lund, The Danish cell project – Part 1: Background and general approach, *Proceedings of IEEE Power Engineering Society General Meeting*, Tampa, FL, 2007, pp. 1–6.

14. D. Pudjianto, C. Rasmsay and G. Strbac, Virtual power plant and system integration of distributed energy resources, *IET Renewable Power Generation*, 1(1), 10–16, March 2007.

15. H. Karimi, J. Davison and R. Iravani, Multivariable servomechanism controller for autonomous operation of a distributed generation unit: Design and performance evaluation, *IEEE Transactions on Power Systems*, 25(2), 853–865, May 2010.

16. R. H. Lasseter, Smart distribution: Coupled microgrids, *Proceedings of the IEEE*, 99(6), 1074–1082, June 2011.

17. W. Cao, J. Wu, N. Jenkins, C. Wang and T. Green, Benefits analysis of soft open points for electrical distribution network operation, *Applied Energy*, 165, 36–47, March 2016.

Fundamentals of Distributed Energy Resources

2.1 Introduction

Distributed energy resources (DER) such as distributed generators, energy storage, electric vehicles and controllable loads are changing the way the power system is developed and operated. Suitably controlled, they can play a multi-functional and flexible role in more effective management of the system. Among the potential benefits claimed for DER are improved system efficiency, reliability, dynamic stability, enhanced power quality, optimized power flow and transmission capacity enhancement. Figure 2.1 shows a typical demand curve and a real-time price regime. It illustrates how energy storage can be used for peak clipping and valley filling. As shown in the figure, surplus electrical energy generated during off-peak hours (e.g. between midnight and 5 a.m., when there is spare generating capacity and the cost of electricity is low) can be stored and subsequently used during peak hours (e.g. between 6 and 9 p.m., when electricity can be sold at a premium price). Electrical energy storage may also be used in combination with intermittent renewable energy sources; for example, the excess electricity from a photovoltaic array can be used to charge a battery during daylight hours for use at night.

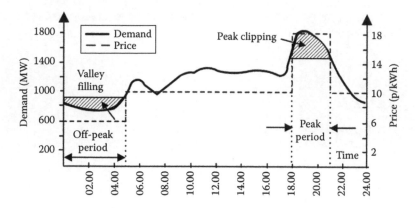

FIGURE 2.1 Benefits of DER.

2.2 Combined Heat and Power Plants

Combined heat and power (CHP) or cogeneration plants allow the simultaneous production of both electricity and heat. Heat, which is an inevitable by-product of generating electricity by burning fuel, can be used directly in an industrial process or for district heating. The capacity of CHP plants varies from a few kW to hundreds of MW depending on the source of energy and the prime mover technology.

Prime movers of different technologies used for CHP are capable of burning a variety of fuels. Table 2.1 summarizes the technologies commonly used for large CHP plants. Small CHP units are based on gas turbines, reciprocating engines or fuel cells and are used for small local generation in houses and commercial buildings and are usually fuelled by natural gas.

TABLE 2.1 CHP Technologies and Their Common Fuels

Technology	Fuel Type
Steam turbine	Any fuel
Single cycle gas turbine	Natural gas, gas oil, landfill gas, LPG or Naphtha
Combined cycle gas turbine	As single cycle gas turbine
Reciprocating engine	Natural gas, gas oil, heavy fuel oil

2.2.1 Steam Turbines

Steam turbines are broadly categorized into three types: fully condensing, back-pressure or non-condensing and pass-out condensing or extraction. Back-pressure and pass-out condensing turbines are used for CHP.

2.2.1.1 Back-Pressure Turbines

In a back-pressure steam turbine, a fuel that may be solid, liquid or gaseous is burned in a boiler, and the resulting high-pressure, high-temperature steam is then passed through a turbine. A set of blades, forming the turbine rotor, are driven by the steam. In this process, the steam expands and its pressure drops. In a back-pressure turbine, all of the energy in the steam that is unused for power generation is exhausted as potentially useful heat. The pressure of the exhaust steam is arranged to be at the value required by the site. This arrangement is shown in Figure 2.2. In the figure, P_e is the electrical power and Q_h is the useful heat. Back-pressure CHP plants are very reliable and have a high availability. CHP plants using back-pressure turbines are available with electrical power outputs of 0.5 MW upwards.

As can be seen from Figure 2.2, the CHP cycle involves many processes. Inside the boiler, a combustion and heat transfer process takes place where high-pressure steam is generated. If the specific enthalpy*

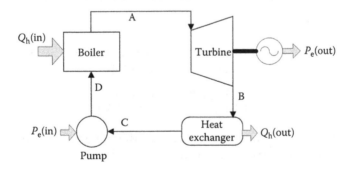

FIGURE 2.2 Elements of back-pressure steam turbine cycle.

---

* Enthalpy is the sum of the internal energy (u) and the product of pressure and volume ($h = u + pV$). The internal energy is the kinetic (translation, vibration and rotation of the molecules) and potential (vibrational and electrical energy of atoms within molecules or crystals) energies of the atoms and molecules. Even though the enthalpy has the same unit as energy, it is not any specific form of energy; it is just a defined variable that is used to simplify calculations in the solution of practical thermodynamic problems.

of the steam at A is h_A and that of the feed water at D is h_D, then disregarding any pressure drops and heat losses, the heat energy added in the boiler is [1]

$$Q_h(\text{in}) = \dot{m}\left[h_A - h_D\right] \tag{2.1}$$

where $\dot{m}$ is the mass flow rate.

Similarly for the other processes, the following equations apply:

$$P_e(\text{out}) = \dot{m}\left[h_A - h_B\right] \tag{2.2}$$

$$Q_e(\text{out}) = \dot{m}\left[h_B - h_C\right] \tag{2.3}$$

$$P_e(\text{in}) = \dot{m}\left[h_D - h_C\right] \tag{2.4}$$

From Equation 2.2, it can be seen that the electrical power output can be increased if the steam input is of high specific enthalpy (h_A), i.e. at high pressure and temperature, and the heat output is of relatively low enthalpy (h_B). If a higher steam pressure is selected at the turbine input, that will increase the capital costs of the boiler and plant running costs. There is a value of steam pressure that results in a given heat/power ratio. The ratio of usable heat to power in a back-pressure steam turbine CHP plant is typically in the range of 3:1 to 10:1 [2]. This makes them unattractive for industries which have a high electricity but low heat demand.

Example 2.1

In the back-pressure steam turbine CHP unit shown in Figure 2.2, the enthalpies at different points are

$h_A = 3000\,\text{kJ/kg};\ h_B = 2500\,\text{kJ/kg};$
$h_C = 500\,\text{kJ/kg}$ and $h_D = 525\,\text{kJ/kg}$

If the mass flow rate of steam through the boiler is 10 kg/s, find the net heat to power ratio, neglecting any losses.

Answer

From (2.2), electrical power output $= 10\times[3000-2500] = 5\,\text{MW}_e$
From (2.3), heat output $= 10\times[2500-500] = 20\,\text{MW}_{th}$
From (2.4), the electrical power consumed by the feed pump $= 10\times[525-500] = 0.25\,\text{MW}_e$
Net electrical power output $= 5 - 0.25 = 4.75\,\text{MW}_e$
$\therefore$ Net heat to power ratio $= 20/4.75 = 4.21$

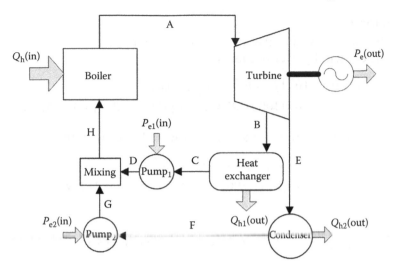

FIGURE 2.3 Elements of pass-out condensing steam turbine cycle.

2.2.1.2 Pass-Out Condensing Steam Turbines

In a pass-out condensing steam turbine, some steam is extracted at an intermediate pressure for the supply of useful heat. The rest of the steam is expanded and cooled in the condenser. This process is shown in Figure 2.3. Q_{h1} is the useful heat output whereas Q_{h2} is the heat rejected by the cycle. Similar to back-pressure steam turbines, in these CHP units for a given fuel input the electric power output decreases with the heat extraction; thus there is an optimum heat-to-power ratio. For a CHP unit of rating less than 5 MW_e operating with a steam export pressure of 15–20 bar, a 1 MW change in electrical output causes, approximately, a 5 MW change in heat output.

Example 2.2

In the pass-out condensing steam turbine CHP plant shown in Figure 2.3, the enthalpies at different points are

$h_A = 3000\,kJ/kg;\quad h_B = 2750\,kJ/kg;\quad h_C = 250\,kJ/kg;$
$h_D = 275\,kJ/kg;\quad h_E = 2250\,kJ/kg;\quad h_F = 250\,kJ/kg$
and $h_G = 275\,kJ/kg$

The mass flow rate of steam through the boiler is 10 kg/s. If 50% of the steam is extracted to the heat exchanger and the remaining 50% condensed, find the heat to power ratio neglecting any losses.

Answer

The mass flow rates B and E are equal at 5 kg/s.
The electrical power output is

$$P_e(\text{out}) = 10 \times \left[h_A - h_B \right] + 5 \times \left[h_B - h_E \right]$$
$$= 10 \times [3000 - 2750] + 5 \times [2500 - 2250] = 3.75\,\text{MW}_e$$

The electrical power consumed by the boiler feed pumps

$$P_e(\text{in}) = 5 \times \left[h_D - h_C \right] + 5 \times \left[h_G - h_F \right]$$
$$= 5 \times [525 - 500] + 5 \times [275 - 250] = 0.25\,\text{MW}_e$$

Heat output

$$Q_h(\text{out}) = 5 \times \left[h_B - h_C \right] + 5 \times [2750 - 250] = 12.5\,\text{MW}_\text{th}$$

Net electrical power output = 3.75 − 0.25 = 3.5 MW$_e$
∴ Net heat to power ratio = 12.5/3.5 = 3.57

2.2.2 Gas Turbines

Gas turbines are widely employed in CHP plants that have power outputs from 1 to 200 MW$_e$. Gaseous or liquid fuels can be used in a gas turbine. The most popular fuel used is natural gas (methane).

Figure 2.4 shows the components of a gas turbine–based CHP plant. Air is first compressed by the compressor. Then heat is added through combusting fuel in the compressed air. The products of combustion at a temperature in the range 900°C–1200°C pass through the power turbine.

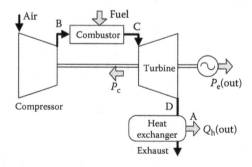

FIGURE 2.4 Simple cycle, single shaft gas turbine.

Inside the power turbine, the gas is expanded to atmospheric pressure rejecting heat at 450°C–550°C.

The heat added by the combustion of fuel (Q_F) is given by [1]

$$Q_F = h_C - h_B \approx \dot{m}c\left[T_C - T_B\right] \qquad (2.5)$$

where
    $\dot{m}$ is the mass flow rate of the air
    c is the specific heat of air
    T is the temperature of the air and gas

The useful heat from the heat exchanger Q_h(out) is given by

$$Q_h(\text{out}) = \dot{m}c\left[T_D - T_A\right] \qquad (2.6)$$

Neglecting losses, the electrical output is given by

$$P_e(\text{out}) = Q_F - Q_h(\text{out}) - P_c \qquad (2.7)$$

where P_c is the power required to run the compressor.

The ratio of heat to power ranges from 1.5:1 to 3:1 [2].

2.2.3 Combined Cycle

In combined cycle systems the heat exhausted from a gas turbine is used to drive a steam turbine. The exhaust heat from the gas turbine allows steam to be generated at high pressure so as to run the steam turbine. Figure 2.5 shows the components of a combined-cycle CHP system. Electric power is generated from generators connected to both the gas and steam turbines. The useful heat is extracted at the exhaust of the steam turbine through a heat exchanger.

2.2.4 Reciprocating Engines

The reciprocating engines used in CHP systems are diesel or Otto engines. In a diesel engine, air is compressed in the cylinder and fuel is injected at high pressure to self-ignite and burn. In Otto engines, a spark plug is used to ignite a pre-mixed charge of air and fuel after compression in the cylinder. CHP units based on diesel engines are available up to ratings of 15 MW$_e$ and can utilize gas oil and heavy fuel oil. Otto engine–based CHP units are available up to 4 MW and usually

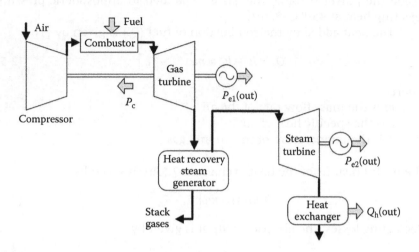

FIGURE 2.5 Combined cycle gas turbine.

operate on gaseous fuel. CHP units are based on existing engine models used for power generation or automotive engines and are less expensive than small gas turbines. The usable heat to power ratio ranges from about 1:1 to 2:1 [3].

The components of CHP based on an internal combustion engine are shown in Figure 2.6. These types of CHP units are mainly used for on-site heat and power generation. Their economics depend on the effective use of the thermal energy in the exhaust gas and cooling system. Heat in the cooling system can produce hot water at a temperature around 80°C–90°C. The engine exhausts gases at around 400°C.

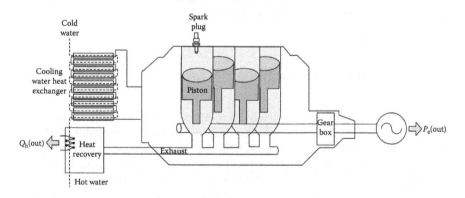

FIGURE 2.6 Components of an Otto engine–based CHP unit.

2.2.5 Micro-Turbines

Micro-turbines are small gas turbines especially designed for domestic and commercial applications. When used as a CHP unit, the control system of the micro-turbine aims to follow the heat load and any deficit or excess of electricity is exchanged with the grid.

Figure 2.7 shows a commonly available micro-turbine. In this unit, the compressed air from the compressor is first heated in a heat exchanger called the recuperator before the fuel is added. This heating process uses the heat in the exhaust air. The rotational speed of the turbine is very high and is in the range of 60,000–100,000 rpm. Two back-to-back electronic power converters are used to obtain 50 Hz of power output.

Micro-turbines with two shafts are also available. In that arrangement, a separate power turbine is used to drive the generator through a gearbox.

Internal combustion engine–based micro-CHP units are also commercially available. Their electrical power rating varies from 1 to 5 kWe. These CHP units generally allow three operation options: heat-led, electricity-led or a combination of the two. The most commonly used mode is heat-led, where the CHP unit is controlled to meet the heating demand in the installation. In an electricity-led operation, the CHP system is controlled to meet the building electrical power demand.

2.2.6 Fuel Cells

Figure 2.8 shows the components of a fuel cell–based CHP unit. The fuel such as H_2 or CH_4 enters the anode and is combined with the ions

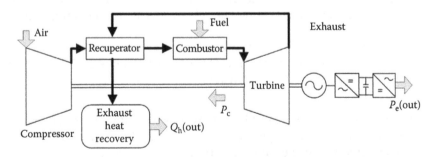

FIGURE 2.7 Components of an internal combustion–based CHP unit. (From U.S. Environmental Protection Agency, Catalog of CHP Technologies, December 2008, available from: https://www.epa.gov/sites/production/files/2015-07/documents/catalog_of_chp_technologies.pdf, accessed on November 19, 2016.)

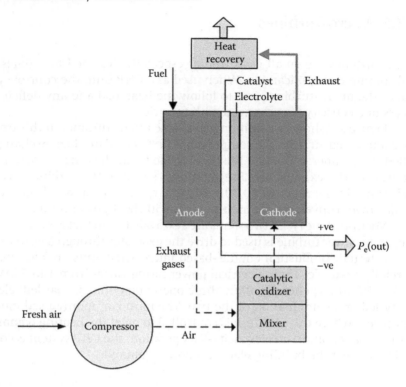

FIGURE 2.8 Fuel cell–based CHP.

exchanged across an electrolyte. Exhaust gas from the electrochemical reaction at the anode is mixed with air and oxidized in a catalytic oxidizer en route to the cathode. At the cathode O_2 from air, exhaust gases from the anode and electrons react and form ions. These ions migrate through the electrolyte to the anode. By-products of this process are water and heat, where water may be released as steam, depending on its temperature.

Fuel cell–based CHP units are of different sizes and types depending on the fuel, electrolyte and membrane used. The following are common fuel cell types for distributed generation:

1. *Proton exchange membrane (PEM)*: This uses a solid polymer electrolyte and operates at low temperature (60°C–80°C). PEM units have a high power density and their power output can be varied.

2. *Alkaline fuel cell (AFC)*: This technology uses alkaline potassium hydroxide as the electrolyte. AFC units operate at low temperature (90°C–250°C).

3. *Molten carbonate fuel cell* (*MCFC*): This uses an alkali metal carbonate as the electrolyte. It operates at 600°C–700°C.

4. *Solid oxide fuel cell* (*SOFC*): This uses a non-porous metal oxide membrane and operates at 750°C–1000°C. However, it is a less mature technology.

Fuel cell–based CHP units are available in different sizes. The largest commercially available CHP unit is MCFC based, with an electrical output of 1.2 MW. The exhaust of this type of CHP unit in the form of a humid flue gas has a high temperature, approximately 300°C–400°C, and a flow rate of 7000 kg/h.

2.3 Photovoltaic Energy Systems

Photovoltaic (PV) energy systems convert energy from the sun directly into electricity. PV systems have been used for many decades mainly as a reliable source of electricity in off-grid applications. However, in the recent past, there has been a rapid growth in grid-connected PV installations in many parts of the world. The reasons for this rapid uptake of solar power are the introduction of government subsidies, decreasing equipment and installation costs, and short installation periods. Meanwhile, the rising cost of grid-supplied electricity is also contributing to PV deployment. PV systems have several advantages, such as extremely low running costs, significant reductions of carbon dioxide (CO_2) emissions, reduction in power losses in transmission and distribution circuits as power is generated at the point of use and reduced imports of fossil fuel.

Silicon-based PV cells are the dominant PV technology to date. Silicon is used in three forms: mono-crystalline, polycrystalline and amorphous. In mono-crystalline PV cells, the crystal is carefully formed so as to obtain a regular structure. Silicon is manufactured in extremely pure form with a single, continuous cubic crystal lattice structure having virtually no defects or impurities. Polycrystalline Si consists of randomly packed 'grains' of mono-crystalline Si. After the controlled manufacture of either mono-crystalline tubular single crystals or polycrystalline blocks, the silicon is cut into thin wafers and fabricated into PV cells. Polycrystalline Si is about 5% less efficient than mono-crystalline Si. Amorphous silicon is utilized in very thin film form to manufacture solar cells. Here, not every Si atom is fully bonded to its neighbouring atoms. In order to neutralize these bonds, 5%–10% hydrogen is incorporated into amorphous silicon.

Alternative technologies are now emerging, although in 2014, out of 39 GWp of PV cell/module shipments, crystalline silicon accounted for 92% [5]. Some of the best known thin film, semi-conductor photovoltaic technologies are based on copper indium diselenide ($CuInSe_2$, or CIS), copper indium gallium diselenide (CIGS) and cadmium telluride (CdTe). Out of these, CdTe shipment in 2014 was 5% of the total PV shipment [5].

A promising, but expensive technology now emerging is multi-junction PV cells. Since Si absorbs light only at wavelengths from 400 to 1100 nm, the light absorption spectrum of a solar cell can be extended by adding other semiconductor materials. For example, Germanium absorbs light in the wavelengths of 900–1600 nm. Multi-junction PV cells consist of more than one semiconductor material and thus absorb light from the whole solar spectrum. Multi-junction PV cells consisting of gallium indium phosphate (GaInP), gallium arsenide (GaAs) and germanium are now reported with efficiencies greater than 45%.

In the previously described PV technologies, the semiconductor acts as both light absorber and the transport medium of the charge carriers. In dye-sensitized cells, these functions are separated – light is absorbed by an inorganic complex – the sensitizer – which is anchored to the surface of a wide band-gap semiconductor. Nanotechnology-based dye-sensitized cells are expected to provide an economic alternative to conventional P–N junction photovoltaic devices in the future.

Organic/polymer solar cells, an emerging technology, have recently attracted wide-scale interest in both the academic and, increasingly, the commercial sector. It is promising in terms of electronic properties, cost and the versatility of functionalization, flexibility and ease of processing.

A variety of PV systems using optical concentrators are also emerging. In some of these, the concentrated solar spectrum is split and each part is directed to a solar cell that has a band gap that matches the spectrum of light incident on it.

The efficiencies of different solar technologies are continuously rising and a good account of their current status can be found in the NREL database [6]. Figure 2.9 shows the trend lines of their efficiencies over the last three decades.

2.3.1 Operation of a PV Cell

A majority of the PV cells in operation are based on silicon and in order to describe the operation of a PV cell, a crystalline silicon structure is

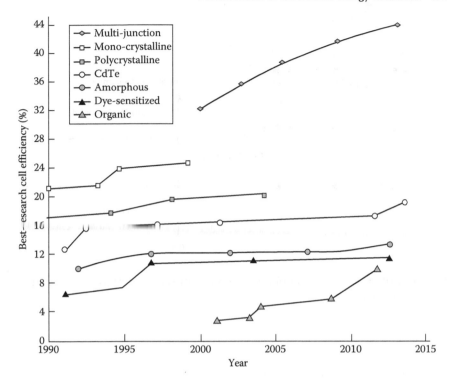

FIGURE 2.9 Trend lines of efficiencies of different PV technologies. (Modified from National Centre for Photovoltaics, available from: http://www.nrel.gov/ncpv.)

used. In an intrinsic Si atom, electrons are bonded together by covalent bonds as shown in Figure 2.10a. The electrons in atoms have different energy levels and they form bands (Figure 2.10b). In the crystalline Si structure, at absolute zero temperature, the valance band is completely filled with electrons and no free electrons exist in the conduction band to allow electrical conduction.

When the temperature increases or the material is struck by a photon, the electrons in the valance band move to the conduction band. The transition of an electron from the valance band to the conduction band leaves a hole in the valance band. Figure 2.11 shows this effect for a single electron moving from the valance band to the conduction band.

Si is not used in its intrinsic state, as the number of current carriers (electrons in the conduction band and holes in the valance band)

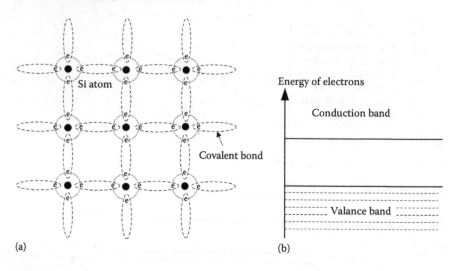

FIGURE 2.10 Intrinsic Si at absolute zero temperature. (a) Crystalline Si structure and (b) energy bands.

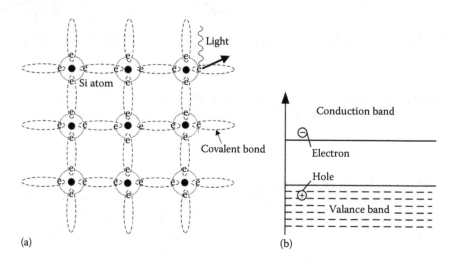

FIGURE 2.11 Intrinsic Si when exposed to light. (a) Crystalline Si structure and (b) energy bands.

is then limited. Normally, impurities from material in Groups 3 and 5 of the periodic tables are added to intrinsic Si. When a donor impurity such as arsenic (As), phosphorous (P), or bismuth (Bi) with five outer electrons is added to the intrinsic semiconductor, the fifth electron in the donor atom does not form a covalent bond and can be easily moved to the conduction band. This is called an N-type semiconductor. Similarly, to form a P-type semiconductor, an acceptor impurity, such as boron (B), gallium (G), or indium (In), with three outer electrons is added to the intrinsic Si.

In a PV cell, a P–N junction is formed by combining a P-type semiconductor with an N-type semiconductor. When the P-type and N-type semiconductors are joined, free electrons from the N-type semiconductor cross the junction and combine with holes near the junction in the P-type semiconductor. At the same time, holes from the P-type region cross the junction and combine with free electrons in the N-type semiconductor. This process produces negative ions in the P-type semiconductor near the junction and positive ions in the N-type semiconductor. The resultant junction region will not have any free charge carriers and is called the depletion region. This region acts like a capacitor having an electric field. The potential barrier created by the electric field prevents further movement of electrons and holes.

When a PV cell is exposed to light, some electrons absorb energy from photons and cross the depletion region. These electrons are collected by the external electrodes and the external circuit. This process is shown in Figure 2.12.

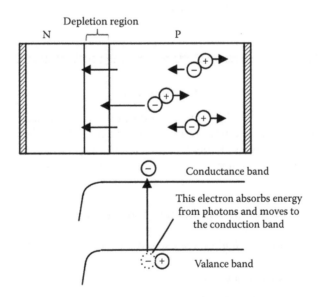

FIGURE 2.12 PN junction.

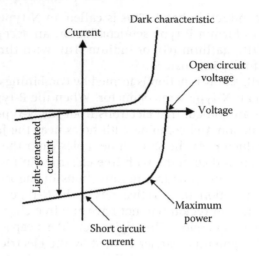

FIGURE 2.13 *I-V* characteristic.

As explained earlier, in dark conditions, no electrons or holes cross the band gap formed by the depletion region so the PV cell acts as a diode. The light-generated current moves the *I-V* characteristics of the PV cell downward, as shown in Figure 2.13.

Example 2.3

A PV cell with sides 120 mm has an open circuit voltage of 0.6 V and a short circuit current of 3 A under an illumination of 1000 W/m². Assuming that the *I-V* characteristic is square, calculate the conversion efficiency at the maximum power point.

Answer

Total power generated at the maximum power point = 0.6 × 3 = 1.8 W

Total solar power incident on the solar cell = 1000 × 0.12 × 0.12 = 14.4 W

Conversion efficiency = 12.5%

2.3.2 Equivalent Circuit of a PV Cell

The bulk resistance of the semiconductor adds a series resistance to the ideal forward-biased diodes. Further partial short-circuiting near the junction introduces a shunt resistance. Therefore the equivalent circuit of a PV cell is a current source, a forward-biased diode, a shunt resistor and a series resistor, as shown in Figure 2.14.

The output current of the PV cell without the shunt and series resistors is given by [7]

$$I = I_L - I_0 \left[e^{\left[qV/nkT \right]} - 1 \right] \tag{2.8}$$

where
I_L is the light-generated current
I_0 is the dark saturation current
V is the output voltage
q is the charge of an electron ($1.60217662 \times 10^{-19}$ C)
k is Boltzmann's constant (1.38×10^{-23} J/K)
T is the absolute temperature
n is a constant between 1 and 2

From Equation 2.8, when the cell is short circuited, i.e. when $V = 0$, the current, $I = I_L$ and is equal to the short circuit current, I_{sc}. Equation 2.8 is then written as

$$I = I_{sc} - I_0 \left[e^{\left[qV/nkT \right]} - 1 \right] \tag{2.9}$$

The I-V characteristic of an ideal PV cell (when the series resistance is zero and the shunt resistance is infinite) is close to a square. In a real PV

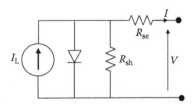

FIGURE 2.14 Equivalent circuit of a PV cell.

---

cell, the output current changes with R_{sh} and R_{se} governed by the following equation:

$$I = I_{sc} - I_0\left[e^{[qV/nkT]} - 1\right] - \frac{V + IR_{se}}{R_{sh}}$$

$$\therefore I = \frac{I_{sc} - I_0\left[e^{[qV/nkT]} - 1\right] - V/R_{sh}}{1 + \left(R_{se}/R_{sh}\right)} \tag{2.10}$$

Example 2.4

Consider a single Si PV cell. At 300 K, and at an irradiance of 1000 W/m², the short circuit current is 2.6 A. Draw the I-V characteristic when (a) $R_{sh} \rightarrow \infty$ and $R_{se} = 0$ and (b) $R_{sh} = 2\,\Omega$ and $R_{se} = 0.1\,\Omega$. Assume $n = 1.25$, and $I_0 = 4$ pA.

Answer

a. From Equation 2.9,

$$\frac{kT}{q} = \frac{1.38 \times 10^{-23} \times 300}{1.60217662 \times 10^{-19}} = 25.85\,\text{mV}$$

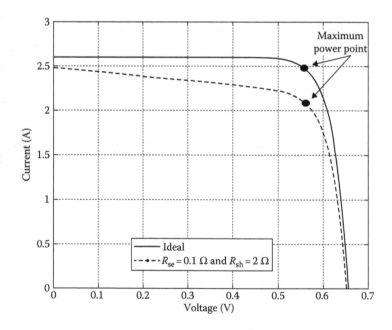

FIGURE 2.15 The effect of series and shunt resistance.

$$\therefore I = I_{sc} - I_0 \left[e^{\left[qV/nkT \right]} - 1 \right] = 2.6 - 4 \times 10^{-9} \left[e^{\left[V/(1.25 \times 25.85 \times 10^{-3}) \right]} - 1 \right]$$

b. From Equation 2.10,

$$I = \frac{\left\{ 2.6 - 4 \times 10^{-9} \left[e^{\left[V/(1.25 \times 25.85 \times 10^{-3}) \right]} - 1 \right] - V/2 \right\}}{1.05}$$

The I-V characteristic for (a) and (b) are shown in Figure 2.15.

2.3.3 Maximum Power Extraction

Using the ideal characteristic shown in Figure 2.15, the power vs. voltage of the solar cell was obtained as shown in Figure 2.16. As can be seen, there is a particular voltage at which the maximum power can be extracted from the PV cell. This point is also marked in Figure 2.16.

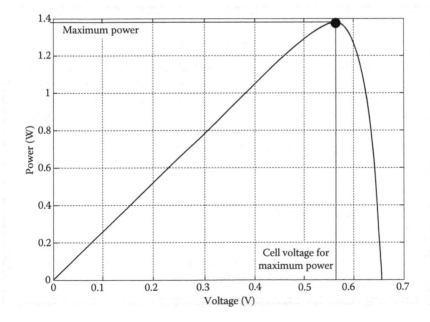

FIGURE 2.16 Power vs. voltage characteristic.

2.3.4 Effect of Irradiance and Temperature

The I-V characteristic of a PV cell varies with irradiance and cell temperature. The short circuit current I_{sc} is proportional to the solar irradiance as shown in Figure 2.17a. The open circuit voltage reduces with the temperature at a rate –2 mV/°C. The effect of temperature on the I-V characteristic is shown in Figure 2.17b.

2.3.5 PV Modules and Arrays

As shown in Figure 2.15, an individual cell produces an open circuit voltage of about 0.6 V. Therefore in practical applications a number of cells are connected in series to obtain a higher open circuit voltage. The short circuit current can be increased by connecting a number of cells in parallel to form a higher power and larger module. The module is the basic building block of a PV array.

Example 2.5

A PV module has 36 cells connected in series. At 25°C, each cell has an open circuit voltage of 0.6 V which reduces with the temperature at a rate of –2 mV/°C. The maximum power point occurs at 80% of the open circuit voltage. If the operating cell temperature is 60°C, at what voltage does the maximum power point occur? For what application is this module suitable?

Answer

Open circuit voltage of a cell at 60°C = $0.6 - 2.0 \times 10^{-3} \times (60 - 25) =$ 0.53 V
Open circuit voltage of the module at 60°C = 0.53 × 36 = 19.08 V
The maximum power point occurs at = 19.08 × 0.8 = 15.26 V
This voltage is suitable for charging a 12 V lead-acid battery.

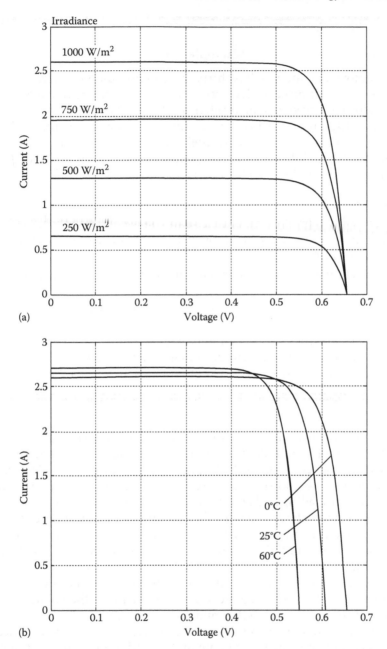

FIGURE 2.17 The effect of series and shunt resistance. (a) Variation with irradiance and (b) variation with cell temperature.

Example 2.6

A photovoltaic module has thirty-six 100 mm diameter circular mono-crystalline cells. Each cell produces a short circuit current of 300 A/m² under solar irradiance of 1000 W/m². In order to form a high power array, four of these modules are connected such that there are two parallel circuits having two series-connected modules. Estimate the open-circuit voltage and short-circuit current of the array at this irradiance?

Answer

Each cell will produce an open-circuit voltage of 0.6 V; therefore, 36 cells in series will give an open-circuit voltage of each series circuit of 21.6 V.

Each cell will produce a short-circuit current of approximately

$$\left[\frac{100}{2}\right]^2 \times \pi \times 10^{-6} \times 300 = 2.36\,A$$

As there are two parallel circuits having two series-connected modules, the array open-circuit voltage will be 21.6 V × 2 = 43.2 V and the array short-circuit current will be 2.36 × 2 = 4.7 A.

In Example 2.5, it was assumed that the cells are identical and the open circuit voltage of the module is equal to the number of cells times the open circuit voltage of one cell. In practice, cells may not be identical. Some cells may have lower short circuit currents due to manufacturing defects, partial shading and aging effects. Then the formation of the total series-connected voltage is due to the combined effect of different cells. This is shown in Figure 2.18, where, in order to show the effect clearly, the mismatch between Cell 1 and Cell 2 is exaggerated. At 2.2 A (the dashed line between A1 and A2), Cell 1 operates at point A1 and Cell 2 operates at point A2. In this case, Cell 1 becomes a load to Cell 2 and the series-connected cells operate at point A. At 1.0 A, the voltage is the addition of the voltages of Cells 1 and 2 (B1 and B2).

The dissipation of power in a poor cell may result in localized heating causing hot spots in the module and material breakdown. When a number of modules are connected in series to form an array, the module with weaker cells consumes power and thus may lead to hot spot formation. In order to avoid this, bypass diodes are normally provided across each module, as shown in Figure 2.19.

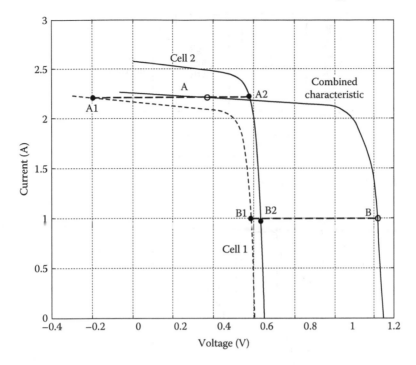

FIGURE 2.18 The effect of connection of non-identical PV cells in series. (From Van Ovrstraeten, R.J. and Mertens, R.P., *Physics, Technology and Use of Photovoltaic*, Adam Hilger Ltd., 1986.)

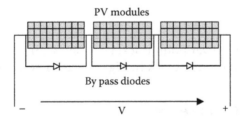

FIGURE 2.19 Connection of a string of PV modules.

2.3.6 Grid-Connected PV Arrays

PV arrays can be operated as a stand-alone unit supplying power to a battery and cluster of loads or as a grid-connected system. The functional components of a grid-connected PV system are shown in Figure 2.20. In some systems, in order to improve the efficiency and flexibility, each

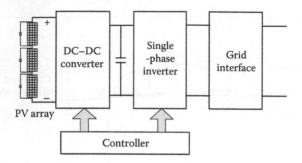

FIGURE 2.20 Grid-connected PV arrays.

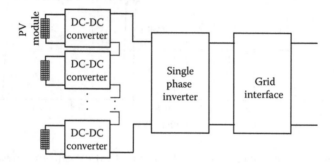

FIGURE 2.21 PV array with a single inverter and DC–DC converter for each module.

PV module is connected to the inverter through its own DC–DC converter (maximum power point tracker), as shown in Figure 2.21.

The function of each block in Figure 2.20 is described next.

2.3.6.1 DC–DC Converter

The function of the DC–DC converter is to extract the maximum power from the PV modules. As discussed in Section 2.3.3, each module should be subjected to a particular voltage to extract maximum power. The inverter maintains a constant voltage across the output of the DC–DC converters. The electronic switches of the DC–DC converter are switched to obtain the required voltage across a module or an array.

As the irradiance of light changes, the voltage across the PV array should be changed to extract maximum power. This is normally achieved by a maximum power point algorithm embedded in the controller of the DC–DC converter. The commonly used algorithms for maximum power point tracking are hill climbing (perturb and observe) and incremental conductance. In the hill climbing technique, the voltage across the module is varied (e.g. reduced) and the power output observed. If the power output increases, then the voltage is reduced by a further step. This is continued until any further decrease in the voltage leads to a reduction in the power output and the voltage change is then reversed.

In the incremental conductance technique, the condition shown in Figure 2.16 that $dP/dV = 0$ at the maximum power point and that $dP/dV = d(IV)/dV = I + V(dI/dV) = 0$ is used to track the maximum power point. The following conditions are used to detect the maximum power point:

At the maximum power point, $dI/dV = -I/V$

To the left of the maximum power point (Figure 2.16), $dI/dV > -I/V$ i.e. $|dI/dV| < I/V$

To the right of the maximum power point, $dI/dV < -I/V$, i.e. $|dI/dV| > I/V$

2.3.6.2 Single-Phase Inverter

A large number of topologies are used for the single-phase inverter [8]. Often the bridge configuration shown in Figure 2.22 is used with pulse

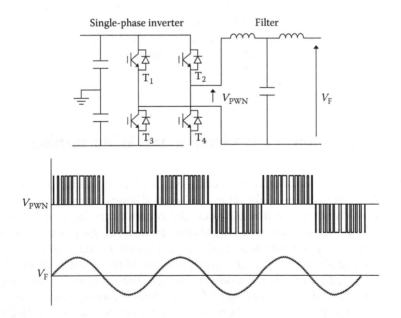

FIGURE 2.22 Single-phase inverter.

width modulation switching. Two complementary insulated gate bipolar transistor (IGBT) pairs, (T_1 and T_4) and (T_2 and T_3), are turned on and off by comparing a sinusoidal 50 Hz signal with a high frequency triangular waveform (usually above 20 kHz to avoid audible noise). The resultant output voltages before filtering (V_{PWM}) and after filtering (V_F) are also shown in Figure 2.22. The inverter can be controlled to maintain a constant DC voltage at the DC–DC converter output and to provide reactive power support to the grid.

2.3.6.3 Grid Interface

The grid interface provides filtering. The filtered output is a sinusoidal with high-frequency noise as shown as V_F in Figure 2.22. In some designs, galvanic isolation is provided at this stage by connecting the inverter through a transformer.

2.4 Wind Energy Systems

Wind energy is one of the mature renewable energy technologies that are added to the network at different voltage levels. The turbine used for wind energy conversion can be of a horizontal axis or vertical axis type. The majority of large wind turbines are horizontal axis, three-bladed types, as shown in Figure 2.23a. Here the nacelle which carries the generator and power conversion stages is placed on the tower. Many different designs of vertical axis wind turbines exist. As shown in Figure 2.23b, some of these include a spiral shape blade running from top to bottom of a vertical shaft.

2.4.1 Operation of a Horizontal Axis Wind Turbine

Figure 2.24 shows a transverse section of the blade element viewed from beyond the tip of the blade. The airstream divides between the upper and lower surfaces. The line between the centre of the aerofoil's leading and trailing edge is known as the 'chord'. In normal operation, it is kept inclined to the airstream at a small angle known as the angle of attack (α). Since the air flow across the upper surface of the aerofoil travels further compared to air over the lower surface, the pressure in the upper surface is reduced, thus creating a lift force. This lift force creates a torque to rotate the turbine blades. There is another force component called the drag force which bends the blade.

(a)   (b)

FIGURE 2.23 Two types of wind turbines. (a) Horizontal axis. (b) Vertical axis.

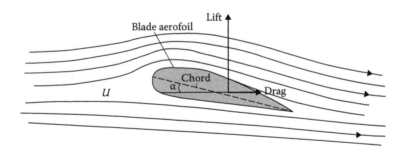

FIGURE 2.24 Flow of air past an aerofoil.

As the turbine rotates the relative direction of wind incident on the aerofoil changes, as shown in Figure 2.25. The angle β is called the pitch angle.

2.4.2 Extraction of Wind Energy from the Turbine

Wind turbines convert the kinetic energy in a stream of air into electricity. If the wind speed is U, then the kinetic energy in a unit mass of

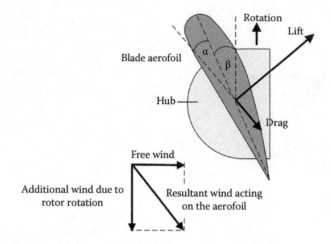

FIGURE 2.25 Relative wind.

air is $(1/2)U^2$. If ρ is the air density in kg/m³ and the area swept by the wind turbine blades is A, the mass flow rate of air is ρAU kg/s. Therefore the total power that is available in the air stream is $(1/2)\rho AU^3$. However only a fraction of this power can be extracted by a wind turbine. This is given by $(1/2)C_p\rho AU^3$, where C_p is the coefficient of performance of the wind turbine. The value of C_p depends on the ratio of the velocity at the rotor tip to wind speed, referred to as the tip speed ratio (λ) and the blade pitch angle (β). Figure 2.26 shows the variation of C_p with λ and β of a typical rotor. As β increases C_p reduces, thus the wind turbine spills energy available in the wind stream.

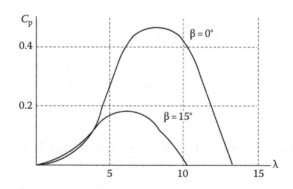

FIGURE 2.26 Variation of power coefficient with λ and β.

Example 2.7

A horizontal axis wind turbine has a rotor diameter of 76 m and reaches its full load output at the generator terminals of 2 MW at a wind speed of 12 m/s. The overall efficiency of the gearbox and generator is 94%. Assume the density of air is 1.2 kg/m³.

a. Calculate the power coefficient (C_p) of the aerodynamic rotor.
b. The turbine operates at a tip speed ratio of 4. Calculate the rotational speed of the aerodynamic rotor.

Answer

a. Area swept by the blades $= \pi \times 76^2/4 = 4536.46 \ m^2$
   Power available in the wind stream $= (1/2)\rho AU^3 = 0.5 \times 1.2 \times 4536.46 \times 12^3 = 4.7 \, MW$
   Mechanical power input to the generator $= 2/0.94 = 2.13 \, MW$
   Therefore the power coefficient $C_p = 2.13/4.7 = 0.45$
b. From $\lambda = \omega r/U$, $4 = \omega \times (76/2)/12$

$$\omega = \frac{12 \times 24}{(76/2)} = 1.26 \, rad/s = 24.06 \, rev/min$$

2.4.3 Wind Turbine Topologies

Depending on the power rating, different wind turbine topologies are used. They can be broadly categorized as fixed-speed wind turbines and variable-speed wind turbines. Figure 2.27 shows a schematic of a fixed-speed wind turbine. A gearbox is connected between the low speed shaft

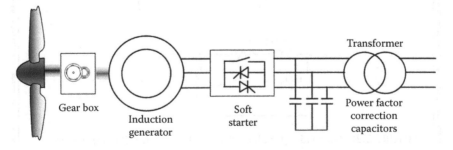

Gear box   Induction generator   Soft starter   Transformer   Power factor correction capacitors

FIGURE 2.27 Component of a fixed speed wind turbine.

and the generator shaft. These turbines usually employ an induction generator. However, in some designs, in order to increase the power extracted
from wind, two induction generators with differing fixed speeds of rotation are used. As an induction generator absorbs reactive power to set
up its magnetic field, a power factor correction capacitor bank is provided. The capacitor bank is normally rated up to the no-load reactive
power requirement of the generator, but this may lead to undesirable
self-excitation if the connection to the network is lost.

Example 2.8

A fixed-speed wind turbine employs a 600 kW, 690 V induction
generator with the following data for the star equivalent circuit:

Stator leakage reactance:	0.08 Ω
Stator resistance:	0.007 Ω (not used in this calculation)
Rotor leakage reactance:	0.09 Ω
Rotor resistance:	0.006 Ω (not used in this calculation)
Magnetizing reactance:	3.0 Ω

Calculate

a. The no-load reactive power drawn by the generator
b. The value of a delta connected capacitor which could supply
0.9 times no-load reactive power of the generator

Answer

a. The equivalent circuit of the wind turbine generator under
no load (when $s = 0$) is given in Figure 2.28. Neglecting the
resistance,

$$\text{No-load current} = \frac{V}{X_{ls} + X_m} = \frac{690/\sqrt{3}}{(3.0 + 0.08)} = 129.34 \text{ A}$$

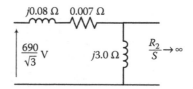

FIGURE 2.28 No-load equivalent circuit of the induction generator.

$$Q = 3VI = 3\left[\frac{690}{\sqrt{3}}\right](129.34) = 154.58 \text{ kvar}$$

b. Since the capacitor bank is connected in delta

$$3 \times (690)^2 \times 100\pi \times C = 154.58 \times 10^3 \times 0.9$$
$$C = 310 \ \mu F$$

As shown in Figure 2.26, for a given wind turbine design there is a particular value of λ at which C_p hence power extraction from the wind becomes maximum. As $\lambda = \omega r / U$, this can only be achieved if the rotor speed of the wind turbine generator is changed with the wind speed. Variable-speed operation not only maximizes the power extraction but also minimizes the mechanical loading on the wind turbine. Figure 2.29 shows the typical operating characteristic of a variable-speed wind turbine. The figure also shows the variation of available power with generator speed for different wind speeds. For rotational speeds less than rated, the generator speed is controlled by a power electronic controller to extract the peak power available. When the maximum rotational speed is reached, the power increases to maintain maximum rotational

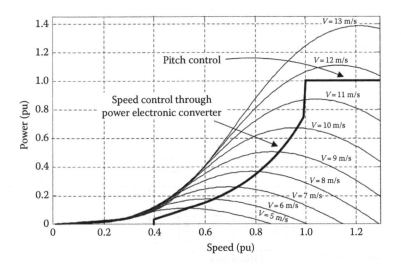

FIGURE 2.29 Operation of a variable-speed wind turbine.

speed. For further increase in wind speeds a part of the available power is spilled using pitch control (where β increases thus C_p reduces).

Some of the common variable-speed generator technologies are described next.

2.4.3.1 Doubly Fed Induction Generator [9,10]

Doubly fed induction generators (DFIG) employ a wound-rotor induction generator with two back-to-back converters feeding the rotor, as shown in Figure 2.30. The mechanical speed of the machine can be controlled by operating the rotor circuit at a variable AC frequency. In this design, the net output power of the machine is a combination of the power coming out of the machine's stator and that from the rotor (through the converter). The phase of the rotor-injected voltage can drive the machine above or below synchronous speed, thus allowing the generator to operate on the maximum power extraction curve shown in Figure 2.29.

2.4.3.2 FSFC-Based Wind Turbines [11,12]

The full scale frequency converter based (FSFC) technology enables the use of synchronous or induction generators for wind turbine applications. They can employ back-to-back voltage source converters (VSC). In some designs with synchronous generators, a diode rectifier and an inverter are used. In rectifier configuration, shown in Figure 2.31, the output of the generator is converted to DC by a rectifier and inverted to the grid frequency (50 or 60 Hz) by an inverter. The chopper is used to boost the voltage and for active power control. In back-to-back converter configuration, as shown in Figure 2.32, the generator-side converter is

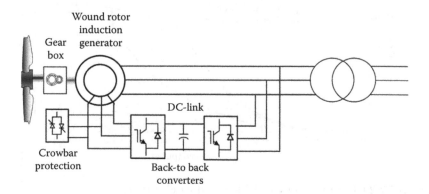

FIGURE 2.30 DFIG wind turbine.

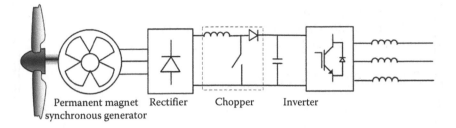

FIGURE 2.31 Gearless FSFC wind turbine with a diode rectifier.

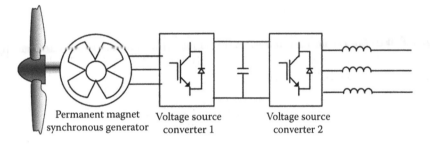

FIGURE 2.32 Gearless FCFS wind turbine with back-to-back converters.

used to control the power extracted from the wind turbine and the grid-side converter is used to maintain the DC link voltage.

The introduction of a large-diameter, multi-pole synchronous generator allows a gearless design. This avoids the mechanical complexity of gears and reduces noise. With gearless designs, typically, the generator has a significantly larger diameter to accommodate a large number of pole pairs (e.g. 84 poles). The generator is directly coupled to the turbine.

Example 2.9

A variable–speed, pitch-regulated wind turbine employs a multi-pole permanent magnet generator with back-to-back voltage source converters. The rated operating conditions of the generator are

No. of poles = 80
Stator terminal voltage = 1 kV line to line rms

Synchronous inductance = 0.5 mH
Shaft torque = 200 kNm
Generator efficiency = 95.5%

The generator windings are star connected.

a. If the generator is working at rated torque at a shaft speed of 100 rev/min, calculate the output power of the generator.
b. Calculate the generator stator current for unity power factor and 0.9 lagging operation.
c. What is the internal voltage of the generator when it is operating at unity power factor?

Answer

a. Output power = $0.955 \times 200 \times 10^3 \times 100 \times 2\pi/60 = 2$ MW
b. $P = \sqrt{3} \times 1000 \times I \times \cos\phi$

$$\text{At unity power factor, } I = \frac{2 \times 10^6}{\sqrt{3} \times 1000} = 1154.7 \text{ A}$$

$$\text{At 0.9 power factor, } I = \frac{2 \times 10^6}{\sqrt{3} \times 1000 \times 0.9} = 1283 \text{ A}$$

c. For the generator, $\omega_e = (p/2)\omega_m = 40 \times 100 \times 2\pi/60 = 418.879$ rad/s

$$X_s = \omega_e L = 418.879 \times 0.5 \times 10^{-30} = 0.2094 \ \Omega$$

$$E = V + IX_s$$
$$= \frac{1000}{\sqrt{3}} + 1154.7 \times j0.2094$$
$$= 577.35 + j242.4$$

Line voltage = 1084.6 V

2.5 Electrical Energy Storage

Although it is difficult to store electricity directly, electrical energy can be stored in other forms, such as potential, chemical, or kinetic energy. Each energy storage technology usually includes a power conversion unit to convert the energy from one form to another. Advanced energy storage technologies based on these principles are emerging as a potential resource in supporting an efficient electricity system. The choice of which storage device should be used can be made mainly considering capacity and energy requirements, footprint of the system, total cost, and environmental conditions. Electrical energy storage technologies can be grouped as electrochemical, mechanical, electrical and chemical technologies.

At the start of the twentieth century, power stations were often shut down overnight, with lead-acid accumulators supplying the residual loads on the direct current networks – one of the first applications of stationary electrical energy storage [13]. Utility companies recognized the importance of the flexibility that energy storage provides in networks, and the first hydroelectric pumped storage system was constructed in Switzerland in 1909, which used separate pump and turbine units [13,14]. According to the recent report, 'Energy Storage Tracker 1Q16', published by Navigant Research [15], 1,119 energy storage projects encompassing 43,305 individual systems were in operation as of 2015. A 6 MW battery storage project with a 10 MWh capacity was recently announced in the United Kingdom. Also, Japan's Hokkaido Electric utility built a 15 MW, 60 MWh battery storage system using vanadium redox flow battery. This will store energy from solar and wind farms to power communities across Hokkaido Island.

Recent improvements in energy storage along with changes in electricity markets indicate that there is an expanding opportunity for electrical energy storage as a cost-effective resource. The market needs for electrical energy storage technology include use of more renewable energy and less fossil fuel, and the future Smart Grid. Electrical energy storage systems have applications such as balancing demand and supply, and grid management in today's power system [16].

Different types of energy storage are suited to different discharge times, from seconds to seasons. Ultra-capacitors, superconducting magnetic energy storage (SMES) and flywheels have short discharge time (seconds to minutes) and offer very rapid power transfer, which can provide frequency regulation and smooth out power fluctuations. Electrochemical energy storage technologies, such as lead-acid (LA), lithium-ion (Li-ion) and sodium–sulphur batteries, have a medium discharge time (minutes to hours) and are useful for power quality and reliability, power balancing and load following, consumer-side time-shifting and generation-side output smoothing. Pumped hydro storage, compressed air energy storage (CAES) and flow batteries have a medium to long discharge time (hours to days) and are primarily used for load following, time shifting and to make renewable energy sources dispatchable. Hydrogen and synthetic natural gas systems have a long discharge time (days to months) and can be useful for seasonal storage.

2.5.1 Electrochemical Storage (Batteries)

Electrochemical batteries are energy storage devices that store and release electricity by alternating between the charge–discharge phases. They can efficiently store electricity in chemicals and reversibly release it according to demand, without harmful emissions or noise, and require little maintenance. A variety of such batteries are available in the market, such as LA, Na-S, nickel–cadmium (NiCd), nickel–metal hydride (NiMh) and Li-ion batteries. Their main advantages are energy density and technological maturity. Their main disadvantage, however, is their relatively low durability for large-amplitude cycling (a few 100 to a few 1000 cycles).

One of the important parameters of a battery is its state of charge (SOC). It is defined as

$$SOC = \frac{\int i\, dt}{Q} \times 100 \tag{2.11}$$

where
  Q is the battery capacity in Ah
  i is the current injected into or drawn from the battery

Example 2.10

A demand profile of a certain island is shown in Figure 2.33. The island has diesel power generating units with a total installed capacity of 800 kW. The average daily load profile of

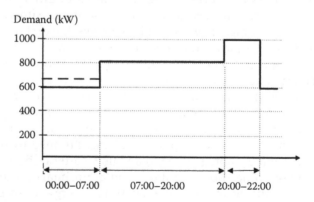

FIGURE 2.33 Demand profile.

---

the island is given in Figure 2.33 (solid line). It was decided to use a battery energy storage unit connected to the island grid supply through an inverter and a 1/33 kV transformer to meet the peak demand.

a. What is the minimum capacity of the battery bank required to supply the peak load without discharging the battery below 20% SOC?
b. If the battery is charged using a constant power charger, draw the modified demand curve to bring the SOC of the battery to 100% with minimum impact to the existing demand curve.

Answer

a. Constant power that the battery should supply from 20:00 to 22:00 is 200 kW.

Therefore, the constant current drawn by the battery

$$= \frac{200 \times 10^3}{\sqrt{3} \times 1 \times 10^3} = 115.47\,\text{A}$$

The minimum capacity is determined by SOC reaching 20% after the battery is discharged at this current rate for 2 h (20:00–22:00 h). Thus from Equation 2.11

$$100 - 20 = 100 \left[\frac{115.47 \times 2}{Q} \right]$$

$$Q = 500\,\text{Ah}$$

b. In order to minimize the impact to the demand curve, the battery should charge from 00:00 to 07:00 h. As the total energy drawn by the battery during the peak is 400 kWh, the power drawn by the charger if constant current charging is used for 7 h

$$= 400/7 = 57.14\,\text{kW}$$

The new demand profile is indicated with a dotted line.

Some of the commonly used battery technologies are described next.

2.5.1.1 Lead Acid (LA)

LA energy storage cells are the most well-known and commercially mature (deployed since about 1890) rechargeable battery technology. A good combination of power density and low price make LA batteries suitable for many applications, such as emergency power supply systems, stand-alone systems with PV, and mitigation of output fluctuations from wind power. Their disadvantages include relatively heavy weight, cycle life limitations and maintenance requirements.

The overall chemical reaction during discharge is [17,18]

$$Pb + PbO_2 + 2H_2SO_4 \rightarrow 2PbSO_4 + 2H_2O$$

The reaction proceeds in the opposite direction during charging.

The largest LA battery installation for grid energy storage is a 14 MW/40 MWh flooded LA system that was built in 1988 in Chino, CA, which is used for load levelling at the Chino substation of Southern California Edison Company [19].

2.5.1.2 Ni-Based Batteries

Ni-metal batteries are another early electrochemical energy storage technology that is used for stationary applications. These batteries all share the same cathode (nickel oxyhydroxide in the charged state) but a different anode that can be cadmium, zinc, hydrogen or a metal-hydride. Nickel–cadmium (NiCd) was in service for many decades. However its place is now taken over by nickel–metal-hydride (NiMh) mainly due to the toxicity of NiCd batteries. Further, NiMh provides 40% higher specific energy than standard NiCd.

2.5.1.3 Li-Ion Batteries

Li-ion batteries, the most important energy storage technology for the portable/consumer electronics market, are also used in hybrid and electric vehicles, as well as for grid storage. Li-ion battery storage systems have high energy density, high voltage/cell (3.2–3.6 V), good charge–discharge characteristics, and low self-discharge; however, their main disadvantage is the high cost due to special packaging and internal overcharge protection circuits [9,10]. Large MW-scale Li-ion battery banks have been in operation for frequency regulation and reserve services since 2008 [17,19].

2.5.1.4 Na-S

Na-S batteries are molten metal batteries based on a high-temperature electrochemical reaction between sodium and sulphur, separated by a beta-alumina ceramic electrolyte, as shown in Figure 2.34. They have a

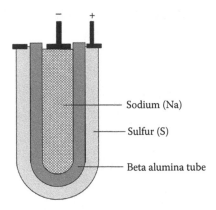

FIGURE 2.34 Na-S battery.

high energy density, high efficiency of charge/discharge, long cycle life and are comparatively inexpensive. Major applications of Na-S batteries include reduction in power fluctuation by load levelling and peak shaving; they are also used to stabilize the intermittency from wind and solar renewable energy generation. The main manufacturer of Na-S batteries is NGK Insulators, Japan, and Na-S battery units have been installed at around 200 sites in Japan, mainly for peak shaving. About 300 MW of installed capacity exists around the world [20].

2.5.2 Flow Batteries

Flow batteries are rechargeable batteries like the secondary batteries discussed in the previous section; however, the energy in flow batteries is stored as charged ions in two separate tanks of electrolytes, one of which stores electrolyte for a positive electrode reaction while the other stores electrolyte for a negative electrode reaction. The power is defined by the size and design of the electrochemical cell and power converter whereas the energy depends on the size of the tanks. As a result, it is possible to scale the energy capacity of the system by increasing the amount of solution in the electrolyte tanks. The major advantage of flow batteries is that power and energy are independent, enabling flexible battery design and they can be used for different applications such as smoothing power output and improving efficiency for renewable energy integration, grid peak shaving, and load levelling.

Vanadium redox batteries (VRBs) are the most mature of all flow battery systems available, in which electrolytes in two loops are separated

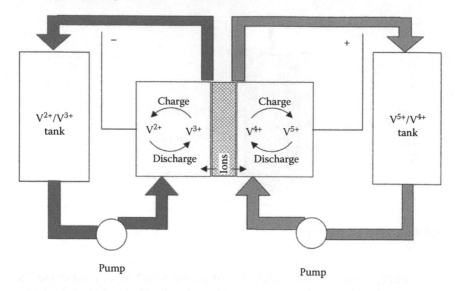

FIGURE 2.35 VRB flow battery system.

by a proton exchange membrane (PEM), as shown in Figure 2.35. The electrolyte is prepared by dissolving vanadium pentoxide (V_2O_5) in sulphuric acid (H_2SO_4). The electrolyte in the positive electrolyte loop contains $VO_2^+-V^{5+}$ and $VO^{2+}-V^{4+}$ ions while the electrolyte in the negative electrolyte loop has V^{3+} and V^{2+} ions. In a VRB cell, the following half-cell reactions are involved [17,19]:

$$\text{Anode: } V^{2+} \leftrightarrow V^{3+} + e^-$$
$$\text{Cathode: } VO_2^+ + 2H^+ + e^- \leftrightarrow VO^{2+} + H_2O$$

The ZnBr battery is a type of hybrid redox flow battery in which two different electrolytes flow past carbon-plastic composite electrodes in two loops separated by a micro-porous polyolefin membrane. As in vanadium redox systems, the ZnBr battery is charged and discharged in a reversible process as the electrolytes are circulated through the appropriate loop. During charging, metallic zinc is plated on the anode and bromide ions are oxidized to bromine and form a polybromide complex with some organic agent in the catholyte. As more zinc is plated across the anode and more polybromide complex is created, the total amount of energy stored in the system increases. During the discharge process, the plated zinc on the anode is oxidized to zinc ion (Zn^{2+}) and bromine is reduced to bromide ion (Br^-) at the cathode. Upon full discharge, both the anolyte and catholyte tanks are returned to a homogeneous aqueous solution of zinc bromide. In a ZnBr battery cell, the following half-cell reactions are involved [17,18]:

$$\text{Anode: } Zn \leftrightarrow Zn^{2+} + 2e^-$$
$$\text{Cathode: } Br_2 + 2e^- \leftrightarrow 2Br^-$$

2.5.3 Mechanical Storage

The pumped hydro system and compression air energy system (CAES) are the two main mechanical energy storage technologies currently employed. For small-scale systems flywheels are considered.

In a flywheel energy storage system, kinetic energy is stored in a rotational cylinder (flywheel) which rotates at very high speed. The kinetic energy stored in a flywheel is given by [21]

$$E = \frac{1}{2}I\omega^2 \text{ J} \qquad (2.12)$$

where $I = mR^2$ is the moment of inertia (m is mass of the flywheel [kg]; R is the radius of gyration [radius at which the total mass is considered to be concentrated]).

A flywheel system has long life, excellent cycle stability, low maintenance, and high energy conversion efficiency. Flywheels are mainly used for short-term energy storage applications such as maintaining power quality and uninterruptible power supply (UPS).

2.5.4 Electrical Storage

2.5.4.1 Ultra-Capacitor

Ultra-capacitors, also known as double-layer capacitors, consist of two conductive metal plates coated with activated carbon and suspended within an electrolyte. A porous, chemically inert dielectric separator physically separates the two plates to prevent short circuit by direct contact. When a voltage potential is applied across the plates, positive ions are collected on one plate while the other gathers negative ions, shown in Figure 2.36. Unlike battery storage units, no chemical or electrochemical reactions take place in ultra-capacitors. In ultra-capacitors, as electrons move from one plate to the other, energy is stored in the electric field.

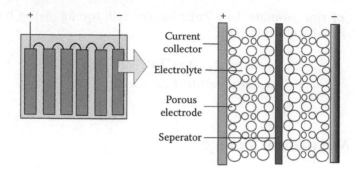

FIGURE 2.36 Ultra-capacitor module and cell. (From Burke, A., *J. Power Sources,* 91, 37, 2000)

The very large surface area and short charge separation distance results in a very high capacitance and the possibility of very fast charge and discharge due to extraordinarily low internal resistance. As the charge and discharge process is a pure physical process which does not change the electrode structure, ultra-capacitors can manage a very large number of cycles and have a long life-span. However, ultra-capacitors are not suitable for the storage of energy over long periods of time, because of their high self-discharge rate, their low energy density and high investment costs. Currently, researchers are experimenting with newer forms of dielectric materials used in ultra-capacitors, to improve capacitance and energy density.

2.5.4.2 Superconducting Magnetic Energy Storage

SMES systems store energy in the magnetic field generated by the flow of direct current in a superconducting coil, of nearly zero resistance, which has been cryogenically cooled to a temperature below its superconducting critical temperature. Once the superconducting coil is charged, the current will not decay and the magnetic energy can be stored indefinitely. The stored energy can be released back to the network by simply discharging the coil. SMES systems recharge within minutes and can repeat the charge/discharge cycle thousands of times without any degradation of the magnet. The other advantages of the system are its high overall round-trip efficiency and very high power output which can be provided for a brief period of time. Moreover, the main parts of an SMES system are motionless, which results in high reliability. SMES systems are currently used only for short duration energy storage because of the energy demand of refrigeration systems and the high cost of superconducting wires.

2.6 Flexible Demand

While supply-side distributed energy resources are increasing, smart appliances and electrical vehicles are emerging as a load that can be shifted and controlled. A number of mechanisms that promote shifting or controlling the load are also emerging. These are generally referred to as demand response (DR) programmes. The U.S. Department of Energy defines DR as 'changes in electric usage by end-use customers from their normal consumption pattern in response to changes in the electricity price over time, or to incentive payments designed to induce lower electricity use at times of high wholesale market prices or when system reliability is jeopardized' [23]. The economic drivers of DR include avoiding the operation of peaking power plants and increasing the reserve margin

Smart appliances such as washing machines and dish washers with 'start delay' function are commercially available [24]. In these appliances, the user can load the appliance at a convenient time and then choose to delay the start of the cycle by a number of hours. This feature allows load shifting from peak to off-peak hours under incentive schemes, such as time of use tariff.

Another smart load that will soon be available in the domestic and commercial sectors is the electric vehicle (EV). An EV may be connected to the future grid at home and at a parking lot near homes or a large number of EVs may be connected at workplaces and shopping malls. EVs can be considered as loads when their battery is charging and that provide energy when their battery discharges. The charging or discharging can be controlled via sophisticated coordination protocols to smooth out fluctuations in power demand, to avoid bottlenecks on the power distribution network and to respond to signals received from the system operators during power system disturbances.

☐ Questions

1. List and explain the advantages and disadvantages of variable speed operation of large wind generators both for the wind turbine and the power system.

2. Draw the elements of a back-pressure steam turbine CHP generator. If the enthalpy of the steam just before the turbine is 3500 kJ/kg, just after the turbine is 2750 kJ/kg and mass flow rate is 10 kg/s, calculate the gross electrical power output of the CHP generator. Assume that

the thermal efficiency of the turbine is 80% and the efficiency of the electrical generator is 95%. If the pump and other auxiliaries consume 5% of the gross output, what is the net electrical power output of the CHP generator?

3. A crystalline silicon PV module has 36 cells connected in series. Each cell is square with sides 150 mm. Assume I_{sc} = 300 A/m² at standard conditions and the cells have a rectangular V–I characteristic (i.e. 100% fill factor)

   a. Estimate the open-circuit voltage, the short circuit current and the power output of the module at an irradiance of 800 W/m².

   b. A solar array is formed by six modules, three in series and two strings in parallel. What is the open-circuit voltage, the short circuit current and the power output of the array at an irradiance of 500 W/m²?

4. The nameplate on a 400 V, 600 kW, 50 Hz, 4 pole induction generator indicates that its speed at rated load is 1510 rev/min. Assume the generator to be operating at rated load. What is the

   a. Slip of the rotor?

   b. Frequency of the rotor current?

   c. Angular velocity of the stator field with respect to the rotor and stator?

5. A country needs a total generation of 12,000 GWh per annum. It was decided to meet 25% of this demand from renewable energy sources. This will be met by P_1 (MW) capacity of wind generation and P_2 (MW) capacity of solar PV with the ratio P_1:P_2 of 5:1.

   a. Assuming that the average capacity factors of the wind generation is 35% and the solar generation is 20%, calculate P_1 and P_2.

   b. If 2 MW wind turbines are used to obtain capacity P_1, how many wind turbines are required? If the rotor radius of each wind turbine is 100 m and the coefficient of performance of each wind turbine is 0.4, calculate the average wind speed required to obtain 2 MW. Assume an air density of 1.25 kg/m³.

☐ References

1. N. Petchers, *Combined Heating, Cooling & Power Handbook: Technologies & Applications*. The Fairmont Press, Inc. & Marcel Dekker, Inc., Lilburn, GA, 2002.
2. Department of Energy and Climate Change, UK, Steam turbines, April 2008. Available from: https://www.gov.uk/government/uploads/system/uploads/attachment_data/file/345189/Part_2_CHP_Technology.pdf. Accessed on November 19, 2016.
3. DUKES-2013, Combined heat and power, August 2013. Available from: https://www.gov.uk/government/uploads/system/uploads/attachment_data/file/279523/DUKES_2013_published_version.pdf. Accessed on November 19, 2016.
4. U.S. Environmental Protection Agency, Catalog of CHP technologies, December 2008. Available from: https://www.epa.gov/sites/production/files/2015-07/documents/catalog_of_chp_technologies.pdf. Accessed on November 19, 2016.
5. P. Mints, Global photovoltaic shipments jump 15% in 2014, SPV Market Research in collaboration with IDTechEx, Feb 2015. Available from: http://www.idtechex.com/research/articles/global-photovoltaic-shipments-jump-15-in-2014-00007454.asp. Accessed on November 19, 2016.
6. Best research-cell efficiencies, National Centre for Photovoltaics. Available from: http://www.nrel.gov/pv/assets/images/efficiency_chart.jpg. Accessed on November 19, 2016.
7. R. J. Van Ovrstraeten and R. P. Mertens, *Physics, Technology and Use of Photovoltaic*. Adam Hilger Ltd., Bristol, UK, 1986.
8. N. Mohan, T. M. Undeland and W. P. Robbins, *Power Electronics – Converters, Applications and Design*, 2nd edn. New York: John Wiley & Sons, Inc., 1995.
9. J. B. Ekanayake, L. Holdsworth and N. Jenkins, Control of DFIG wind turbines, *Power Engineer*, 17(1), 28–32, February 2003.
10. F. M. Hughes, O. Anaya-Lara, N. Jenkins and G. Strbac, Control of DFIG-based wind generation for power network support, *IEEE Transactions on Power Systems*, 20(4), 1958–1966, 2015.
11. B. Fox, D. Flynn, L. Bryans, N. Jenkins, D. Milborrow, M. O'Malley, R. Watson and O. Anaya-Lara, *Wind Power Integration: Connection and System Operational Aspects*, IET Power and Energy Series, Vol. 50, IET, UK, 2007.
12. V. Akhmatov, A. H. Nielsen and J. K. Pedersen, Variable-speed wind turbines with multi-pole synchronous permanent magnet generators. Part I: Modelling in dynamic simulation tools, *Wind Engineering*, 27, 531–548, 2003.
13. J. N. Baker and A. Collinson, Electrical energy storage at the turn of the millennium, *Power Engineering Journal*, 13(3), 107–112, June 1999.
14. J. A. Suul, Variable speed pumped storage hydropower plants for integration of wind power in isolated power systems. Available from: http://cdn.intechopen.com/pdfs/9345/InTech-Variable_speed_pumped_storage_hydropower_plants_for_integration_of_wind_power_in_isolated_power_systems.pdf. Accessed on November 19, 2016.

15. Navigant Research, Energy storage tracker 1Q16. Market share data, industry trends, market analysis, and project tracking by world region, technology, application, and market segment, March 2016. Available from: https://www.navigantresearch.com/research/energy-storage-tracker-1q16. Accessed on November 19, 2016.

16. DG ENER working paper. The future role and challenges of Energy Storage, DG ENER working paper, January 2013. Available from: http://ec.europa.eu/energy/index_en.htm. Accessed on November 19, 2016.

17. D. Rastler, Electrical energy storage technology options, EPRI, December 2010. Available from: http://www.epri.com/search/Pages/results.aspx?k=Electricity%20Energy%20Storage%20Technology%20Options. Accessed on November 19, 2016.

18. R. Liu, L. Zhang, X. Sun, H. Liu and J. Zhang, *Electrochemical Technologies for Energy Storage and Conversion*, Vols. 1 and 2. Wiley-VCH Verlag GmbH & Co., Germany, January 2012.

19. D. H. Doughty, P. C. Butler, A. A. Akhil, N. H. Clark and J. D. Boyes, Batteries for large-scale stationary electrical energy storage, December 2013. Available from: http://www.electrochem.org/dl/interface/fal/fal10/fal10_p049-053.pdf. Accessed on November 19, 2016.

20. D. Anthony and T. Zheng, Is sodium sulfur (NaS) battery a viable grid energy storage solution? December 2013. Available from: http://www.cleantechblog.com/blog/2012/01/is-sodium-sulfur-nas-battery-a-viable-grid-energy-storage-solution.html. Accessed on November 19, 2016.

21. A. Ter-Gazarian, *Energy Storage for Power Systems*, IEE Energy Series, Vol. 6. London, UK: Peter Penguins Limited, 1994.

22. B. Andrew, Ultracapacitors: Why, how, and where is the technology, *Journal of Power Sources*, 91, 37–50, 2000.

23. Benefits of demand response in electricity markets and recommendations for achieving them, A report to the United States Congress Pursuant to section 1252 of the energy policy act of 2005, February 2006. Available from: http://emp.lbl.gov/sites/all/files/ REPORT%20lbnl%20-1252d.pdf. Accessed on November 19, 2016.

24. M. Presutto, R. Stamminger, R. Scialdoni, W. Mebane and R. Esposito, Preparatory studies for eco-design requirements of EuPs: Domestic washing machines and dishwashers, December 2007. Available from: http://www.ebpg.bam.de/de/ebpg_medien/014_studyf_08-12_part6-7.pdf. Accessed on November 19, 2016.

Management of Distributed Energy Resources

3.1 Introduction

Existing electrical energy systems were designed and built to accommodate large power-generating plants and to serve demand that traditionally has been viewed as uncontrollable and inflexible. The operation and management of the power system was centrally controlled. In the past decades, there has been a revival of interest in connecting distributed energy resources (DERs), such as distributed generation and energy storage, to the distribution network and in microgeneration and flexible loads at the premises of end users. The current policy of installing DERs encourages connection rather than integration. Typically, DERs have been installed with a 'fit and forget' approach, based on the legacy of a passive distribution network. Under this regime, DERs are not visible to the system operator. While it can displace energy produced by centralized generation, distributed generation installed as 'fit and forget' does not necessarily provide capacity at times it is most needed. This approach leads to various technical issues that need to be solved, for example undesirable voltage excursions, increase in peak current, increase in network fault levels and reduced power quality.

For effective operation of distribution networks under constrained operating conditions, various approaches have been proposed for better management of DERs. Demand-side integration (DSI) is a set of measures to use loads to support network operation/management and improve the quality of power supply. In practice, the potential of DSI depends on the availability and timing of information provided to consumers, the duration and timing of their demand response, performance of the ICT infrastructure, metering, automation of end-use equipment and pricing/contracts. Another approach is to break a distribution network into smaller entities such as microgrids and unit control areas (CELLs). The microgrid concept assumes a cluster of loads and microsources operating as a single, controllable system that provides both power and heat to its local area. A CELL is used to define a wider area of network in which a collection of DERs can be controlled in response to a number of objectives and can be considered as an entity that consists of a number of microgrids. Its electrical capabilities are defined as a black box and this encourages a hierarchical structure in which CELLs can be further aggregated or subdivided. The concept of the virtual power plant (VPP) is to aggregate many small generators into blocks that can be controlled by the system operator, whose energy output can then be traded. Aggregating the DER units into a portfolio makes them visible to the system operator and can be controlled actively. The aggregated output of the VPP is configured to have technical and commercial characteristics similar to a central generation unit.

3.2 Demand-Side Integration

Load control or load management has been widespread in power system operation for a long time with a variety of terminology used to describe it. The term demand-side management (DSM) was used in the 1970s for a systematic way of managing loads. Later on demand response (DR), demand-side response (DSR), demand-side bidding (DSB) and demand bidding (DB) came to be used to describe a variety of different demand-side initiatives. To avoid the confusion caused by such overlapping concepts and terminologies, as recommended by the International Council on Large Electric Systems (CIGRE), 'demand-side integration' is used in this chapter to refer to all aspects of the relationships between the electric power system, energy supply, and end-user load.

3.2.1 Services Provided by DSI

DSI can provide various services to the power system by modifying load consumption patterns. Such services include load shifting, valley filling, peak clipping, dynamic energy management, energy efficiency improvement and strategic load growth. Simple, daily domestic load profiles are used to illustrate the function of each service, as shown in Figures 3.1 through 3.4.

Load shifting is the movement of load between times of day (from on-peak to off-peak) or seasons. In Figure 3.1, a load, such as a wet appliance (washing machine), that consumes 1 kW for 2 h is shifted to off-peak time.

Figure 3.2 shows the main purpose of valley filling, which is to increase off-peak demand through storing energy, for example in a battery of a plug-in electric vehicle or thermal storage in an electric storage heater. The main difference between valley filling and load shifting is that valley filling introduces new loads to off-peak time periods, but load shifting only shifts loads so the total energy consumption is unchanged (as shown in Figure 3.1).

Peak clipping reduces the peak load demand, especially when demand approaches the limits of feeders/transformers, or the supply limits

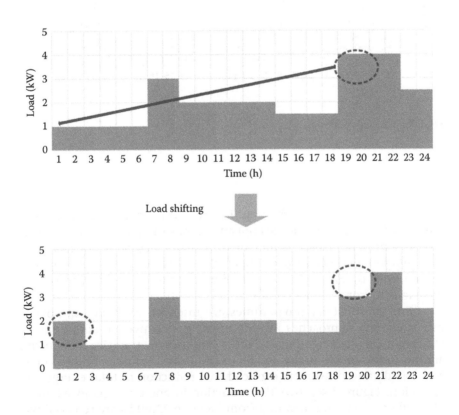

FIGURE 3.1 Load shifting.

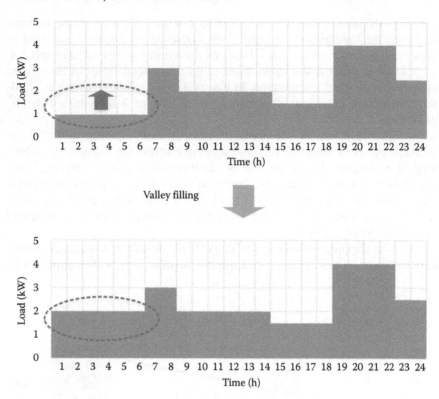

FIGURE 3.2 Valley filling.

of the entire system. Peak clipping (Figure 3.3) is primarily done through direct load control of domestic appliances, for example reducing the thermostat setting of a space heater or controlling electric water heaters or air-conditioning units. As peak clipping reduces the energy consumed by certain loads (in Figure 3.3, 2 kWh of load is curtailed), often consumers have to compromise their comfort.

Energy efficiency programmes are intended to reduce the overall use of energy. Approaches include offering incentives to adopt energy-efficient appliances, lighting and other end uses or strategies that encourage more efficient electricity use, for example the feedback of consumption and cost data to consumers can lead to a reduction in total energy consumption. Figure 3.4 shows the reduction in energy demand when ten 60 W filament lamps (operating from 18.00 to 22.00 h) are replaced with 20 W compact fluorescent lamps.

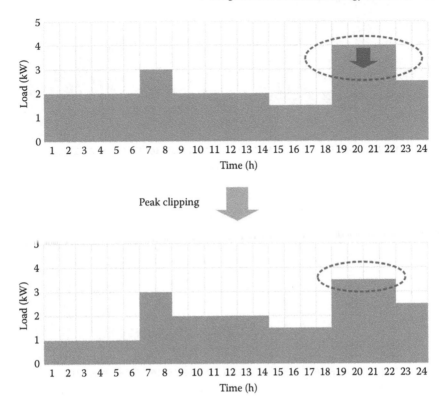

FIGURE 3.3 Peak clipping.

3.2.2 System Support from DSI

Emergency load shedding has been used in many power systems to maintain their integrity during major disturbances, often triggered by under-frequency relays when the frequency drops below a certain threshold, for example 48.8 Hz in England and Wales, and causes tripping of entire distribution feeders. Load shedding is planned by the transmission system operator (TSO) but is implemented by the distribution network operators (DNOs) who arrange the tripping of distribution feeders and choose which feeders are to be tripped.

During normal operation, the GB TSO (NGET) maintains the frequency at 50 ± 0.2 Hz. In order to maintain frequency, NGET procures frequency response services. When the frequency increases, this high frequency response is used to reduce the power output of the large generators and hence the frequency. On the other hand, a sudden drop in frequency is contained by using primary response (Figure 3.5).

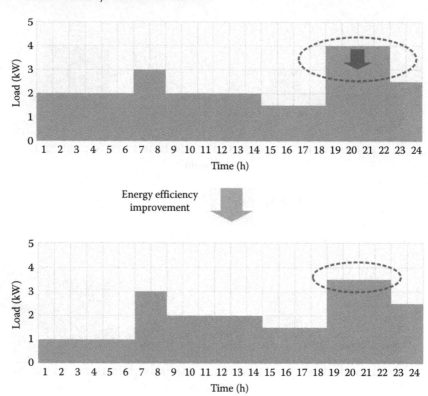

FIGURE 3.4 Energy efficiency improvement.

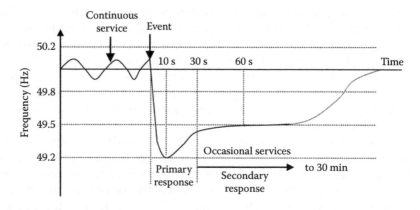

FIGURE 3.5 Frequency control in England and Wales.

This should be delivered within 10 s and maintained for another 20 s. The system frequency is brought back to normal using secondary response which starts within 30 s and lasts for 30 min. If the frequency continues to drop below 48.8 Hz, load shedding (i.e. disconnecting demand) is used to prevent shutdown of the power system.

Primary and secondary response is usually provided by partially loaded generators increasing their output. However, this incurs significant costs as de-loading a steam turbine set by 100 MW typically only provides around 20 MW of primary response. Primary response requires energy to be used that is stored as high-pressure steam in the boiler drum; therefore, only a fraction of the de-loaded power can be used. Operating a steam turbine set below its maximum output reduces its thermal efficiency.

DSI programmes can significantly reduce the requirement for primary and secondary response from partially loaded generators by shedding load in a controlled manner. Large loads that are contracted to provide frequency response are typically steel works or aluminium smelters, although hospitals and banks that have their own generators can also provide this service. Using load in this way reduces system operating cost and, depending on the alternative generation being used, CO_2 emissions.

Similarly, DSI can support distribution network operation. Distribution circuits are planned and designed with spare capacity to account for future demand growth and contingency overloads. With the electrification of the transport and heating sectors, the load anticipated to be connected to the distribution network will require the uprating of network components. Conventionally, the capacity of distribution networks is estimated considering temporal diversity of appliance use since individual appliances are turned on and off at different times. The smart metering infrastructure will enable more complex control signals to be sent to smart appliances that will result in dynamic changes in consumption and could reduce the diversity factor due to the synchronizing effect of smart control of appliances. DSI is able to coordinate such smart appliances so provide support to distribution network operation.

Case Study 3.1

A DSI regional controller is introduced to implement network support from smart appliances. Washing machines, dishwashers and tumble dryers are considered in this case study. The proposed regional controller is illustrated in Figure 3.6. The controller is located at the secondary substation. The controller will schedule the operation of the smart appliances that are connected to the

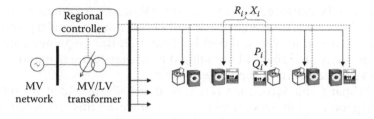

FIGURE 3.6 Concept of a DSI regional controller for smart appliances.

electricity distribution network downstream of the controller's location.

The objective of the DSI algorithm is to minimize the cost of operating the smart appliances. The controller must ensure that the aggregated demand of appliances does not drive the network parameters outside operational limits. Therefore network constraints, both network thermal limits and network voltage limits, are considered in the shifting algorithm.

The thermal constraint, modelled using Equation 3.1, keeps the loading at the secondary substation transformer lower than the transformer rating.

$$\sum_{k=1}^{K} \sqrt{\left(P_{Dk}^t + P_{Ak}^t\right)^2 + \left(Q_{Dk}^t + Q_{Ak}^t\right)^2} \leq S_{\text{transformer}}, \quad \forall t \in \left(t_0, \ldots, t_0 + T\right)$$

(3.1)

where
$S_{\text{transformer}}$ is the MV/LV transformer rating
k is the index of the residential consumer
K is the number of residential consumers connected to the LV network
P_{Ak}^t, Q_{Ak}^t is the power consumption from controllable appliances for customer k at time t
P_{Dk}^t, Q_{Dk}^t is the power consumption from other household appliances for customer k at time t
T is the optimization window
t_0 is the current time step

The voltage constraints keep the voltage across the LV feeder within the declared voltage range of 230 V + 10% and −6%. Because only loads are connected to the test network, it is sufficient to consider the voltage constraints at only two nodes:

the start and the end of the feeder. The nodal voltage constraints considered are based on the approximation $\Delta V = PR + XQ$.

The first constraint, modelled in Equation 3.2, ensures the voltage at the customer connected farthest from the substation is above the lower bound of the voltage range. The second constraint, modelled in Equation 3.3, ensures the voltage at the customer connected closest to the substation is below the upper bound of the voltage range.

$V_{\text{End\_Feeder}} \geq V_L$

$$V_S - \left[\sum_{i=1}^{n} \left(R_i \cdot \sum_{j=i}^{n} \left(P_{Dj}^t + P_{Aj}^t \right) \right) + \sum_{i=1}^{n} \left(X_i \cdot \sum_{i=i}^{n} \left(Q_{Dj}^t + Q_{Aj}^t \right) \right) \right] / V_S \geq V_L,$$

$$\forall t \in \left(t_0, \ldots, t_0 + T \right) \tag{3.2}$$

$V_{\text{Start\_Feeder}} \leq V_U$

$$V_S - \left[R_1 \cdot \sum_{j=1}^{n} \left(P_{Dj}^t + P_{Aj}^t \right) + X_1 \cdot \sum_{j=1}^{n} \left(Q_{Dj}^t + Q_{Aj}^t \right) \right] / V_S \leq V_U,$$

$$\forall t \in \left(t_0, \ldots, t_0 + T \right) \tag{3.3}$$

where

V_L, V_U are the lower and upper voltage limits of the LV network

V_S is the voltage of the LV busbar at the secondary substation

i is the index of the cable section across the LV feeder

n is the number of cable sections on the LV feeder

R_i is the resistance of the cable section i

X_i is the reactance of the cable section i

P_{Aj}^t, P_{Dj}^t is the real power consumption connected to the cable section j at time t

Q_{Aj}^t, Q_{Dj}^t is the reactive power consumption connected to the cable section j at time t

Figure 3.7 shows the transformer loading with different uptake rates of smart appliances without the regional DSI controller. The smart appliances reacting autonomously to the Time of Use tariff decrease the load diversity (which shows the difference between the peak of coincident and non-coincident demands of two or

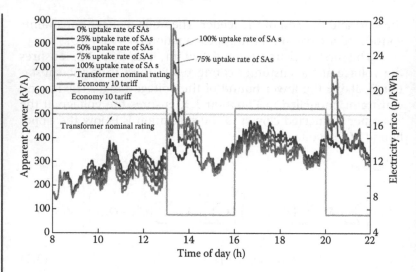

FIGURE 3.7 Loading of substation transformer without a regional DSI controller (SAs denotes Smart Appliances).

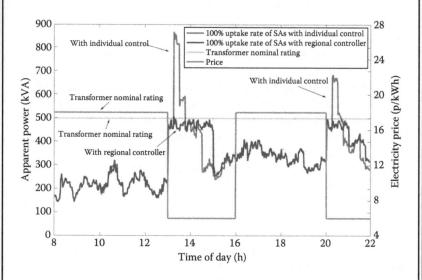

FIGURE 3.8 Loading of substation transformer with a regional DSI controller (smart appliances with individual control will only minimize its own cost without considering the system impact).

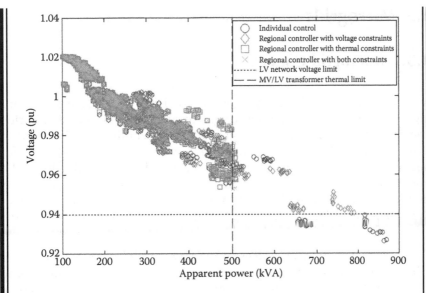

FIGURE 3.9 Evaluation of different control strategies for smart appliances by observing the transformer loading and voltage at the farthest customer from the substation over the time span of a day.

more individual loads) at the off-peak hours of the tariff. The loading of the transformer exceeds its rated capacity.

Figure 3.8 shows that with the regional controller, the demand recorded at the transformer remains below the transformer rating.

The performance of the regional controller considering various constraints is compared in Figure 3.9. Each point characterizes the voltage recorded at the customer farthest from the substation and the corresponding demand on the MV/LV transformer. The regional controller considering both thermal/voltage constraints behaves similarly to the case when only the thermal constraint is enabled. The points around the thermal rating (500 kVA) are still above the voltage limit (0.94 pu), while the points around the voltage limit correspond to a much higher demand than the transformer rating.

This shows that for the test network, the thermal rating of the MV/LV transformer is the most limiting constraint in the operation of appliances on the network and that DSI is able to manage these appliances in order to keep the distribution network operating within limits.

3.3 Microgrids

3.3.1 Microgrid Concept

A microgrid is a small electricity generation and distribution system containing distributed generation, energy storage systems, loads and monitoring and protection devices. It is an autonomous system that is self-controlled and self-managed. An energy microgrid provides users thermal energy for heating and cooling in addition to electricity. A fundamental feature of a microgrid is that it can operate either in grid-connected or islanded mode. In the grid-connected mode, the microgrid exchanges electrical energy with the bulk power grid. The advantages of microgrids include the following:

1. The controllable power sources and energy storage systems in a microgrid can accommodate the fluctuations of renewable power generation and thus improve power quality. Diverse operational objectives such as minimizing operational costs and maximizing energy utilization can be achieved through effective management of distributed generation and loads [1] and the design of energy management systems[2].

2. A microgrid can provide differentiated custom-made services to satisfy all kinds of loads. For example, a microgrid can supply important loads with high-reliability power, while supplying less important loads with cheaper power at lower reliability.

3. In distribution networks, a microgrid acts as a virtual power source or load. Therefore, peak shaving can be realized through coordinated control of distributed generation and loads. Moreover, the adverse impacts of a high penetration of distributed generation can be reduced to a great extent by microgrids, thus helping system operators manage the distribution network more easily.

4. Because of the ability of independent operation, microgrids assist distribution networks with self-healing after faults. When there are faults in the network, many microgrids can keep supplying important loads.

5. A microgrid can be constructed and operated by power users, power companies or independent third-party energy companies. This kind of multi-party operation encourages all kinds of stakeholders to participate in the construction of renewable power generation facilities, thus promoting the revolution of the market model and mechanism in the energy field.

3.3.2 Microgrid Structure

3.3.2.1 AC Microgrid

In an AC microgrid, distributed generators and energy storage systems are connected to an AC bus through power electronics devices, as shown in Figure 3.10. Through on/off control at the point of connection (PC), the microgrid can be switched into either grid-connected mode or islanded mode.

3.3.2.2 DC Microgrid

A DC microgrid has a DC bus to which distributed generators, energy storage systems and loads are connected. The DC network is connected to the bulk AC power grid through a power electronics inverter, as shown in Figure 3.11. AC and DC loads at different voltage levels can be supplied by a DC microgrid through power electronic devices. The fluctuation

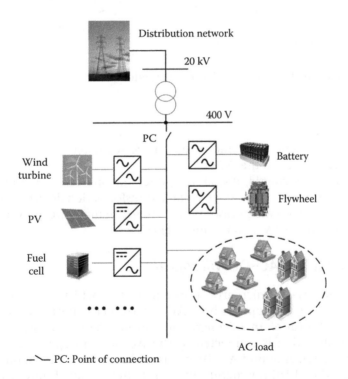

FIGURE 3.10 Typical structure of an AC microgrid.

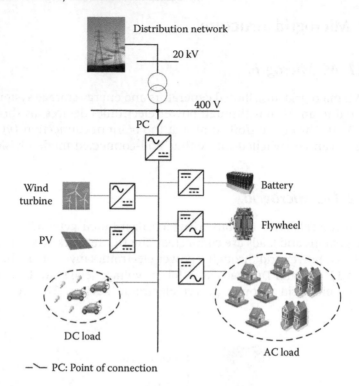

FIGURE 3.11 Typical structure of a DC microgrid.

of distributed generation and loads can be regulated by energy storage systems connected to the DC bus.

In a DC microgrid, distributed generators are connected to the DC bus only through a single-stage voltage transformation device. This structure is more economic in cases where there are many DC power sources and loads, such as PV systems and fuel cells.

3.3.2.3 Hybrid AC–DC Microgrid

The hybrid AC–DC microgrid shown in Figure 3.12 is composed of an AC and DC bus and supplies both AC and DC loads. It can be considered as a special form of an AC microgrid considering the DC network as a power source that is connected to the AC bus through a power electronic inverter. A hybrid AC–DC microgrid combines the characteristics of both AC and DC microgrids and can better supply different types of loads.

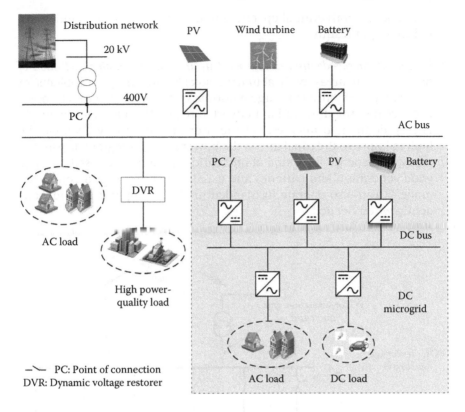

FIGURE 3.12 Typical structure of a hybrid AC–DC microgrid.

3.3.3 Microgrid Applications

As a system that provides users with custom power supply services, a microgrid can be applied to various scenarios with different components, structures and operational characteristics, including

1. *Independent microgrids on islands or in remote areas*: It is difficult and expensive to construct a conventional power grid on islands or in remote areas, so a microgrid can be attractive. For a microgrid in an island or remote area [3], the components and structures should be decided based on local environmental conditions. For example, when there are abundant wind and solar resources, the microgrid can be composed of wind turbine generators, PV arrays, diesel generators and batteries. This kind of microgrid then operates in islanded mode. Although diesel generators are still needed considering the uncertainty of wind and

solar energy, both annual operating hours and diesel consumption are reduced significantly.

2. *Renewable energy–dominated microgrid in areas with a high penetration of renewables*: In areas with abundant solar/wind energy, problems of overvoltage may occur if large amounts of solar/wind generation are connected to the distribution network directly. Therefore, a renewable energy–dominated microgrid at the user or community level can be built to improve the capability of the power grid to integrate distributed renewable energy. This kind of microgrid is mainly composed of solar/wind generation and batteries and usually operates in grid-connected mode. It can also operate in islanded mode to supply users independently when needed.

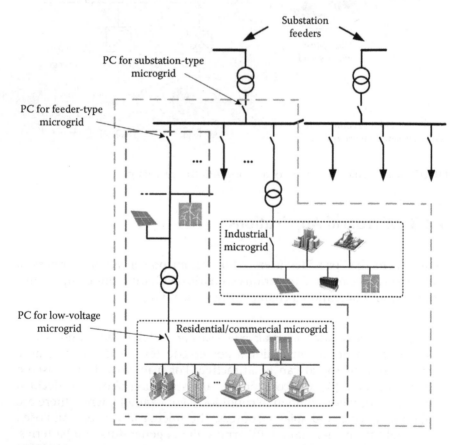

FIGURE 3.13 Typical structure of a highly integrated microgrid in the distribution network.

3. *Integrated energy microgrid in areas with diverse energy sources and demands*: When there are abundant energy sources and diverse demands like cooling/heating/electricity demands, an integrated energy microgrid can be constructed to serve large public buildings, schools or hospitals [4]. The main purposes of this kind of microgrid are to integrate building/ community energy-saving technology, improve integrated energy utilization and realize efficient use of energy. In such microgrids, electrical energy is generated by solar, wind, geothermal or biomass energy and is stored in electrical or thermal energy storage systems. In addition, combined cooling, heating and power is a typical characteristic of such microgrids.

4. *Highly integrated microgrid in distribution networks with extensive distributed energy*: This kind of microgrid is an important part of smart distribu tion networks, as shown in Figure 3.13. Because distributed generators like PV systems and wind generators that are directly connected to the distribution network cannot supply loads independently, they have to be shut down when there are faults in the distribution work. By integrating distributed generators into branch-level, feeder-level and substation-level microgrids in distribution networks, distributed generators can supply key loads when there are faults in the bulk power grid during disasters and can improve the self-healing ability of distribution networks.

Case Study 3.2

A photovoltaic energy storage system microgrid in a building is described in this section. It is a grid-connected AC microgrid located in Tianjin Eco-City. Poly-crystalline silicon PV modules with a total rating of 300 kWp are installed on the roof, walls and carport of the building. Moreover, 648 kWh of lithium batteries and 100 kWmin supercapacitors are installed in order to reduce the influence of PV power fluctuation and supply key building loads when a fault occurs in the external power grid. The mixed energy storage system makes the best of the advantages of different energy storage systems, and the impact of microgrid mode switching can be softened.

The connection of the main devices in the microgrid is shown in Figure 3.14. It is a low-voltage microgrid, and the central control system coordinates the operation of all the devices. The microgrid can operate in seven different modes, as listed in Table 3.1. In normal operation, the system does not feed power

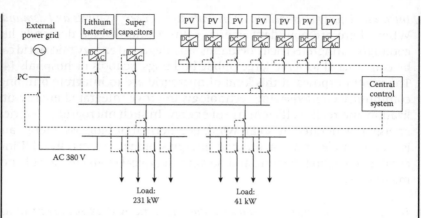

FIGURE 3.14 Structure of a microgrid.

back to the external power grid. When there is surplus PV power, the power is delivered to the energy storage system, or curtailed, depending on the state of charge (SOC) of the storage systems. When the PV power is insufficient (e.g. due to clouds during daytime, or in the evening), the demand is met from the energy storage system or the external power grid. The central control system monitors the power exchange at the point of connection and regulates the charging/discharging of the battery system. When there are faults in the external power grid, the microgrid will be switched to islanded mode, in which the mixed energy storage system (lithium batteries and supercapacitors) operates automatically to maintain the system voltage and frequency, according to the available PV power and demand.

The islanded operation of the microgrid on a typical day is illustrated in Figure 3.15. In the evening, the demand relies entirely on the energy storage systems. In the morning, the PV power keeps increasing as solar radiation grows. As a result, the energy storage system switches from discharging mode to charging mode until its SOC reaches the upper limit. If the PV power is still higher than the demand at this time, some PV modules will be cut off. In the afternoon, the energy storage system switches back to the discharging mode with the decrease in the PV power output. When the PV power reaches zero, all the load demand of the building will be supplied by the energy storage system. In Figure 3.15, the PV curve shows the available PV power during the day, while the PV_T curve represents actual usage of PV power by the system.

TABLE 3.1 Main Operating Modes of a Microgrid

Mode		Time	Load Level	Net Load (Load and PV)	Battery Mode	PV Mode	Power from External Grid
1	Grid-connected	Night	Low	Positive	Discharging	—	No
2			High	Positive	Discharging	—	Yes
3		Day	Low/High	Positive	Discharging	All in	Yes/No
4				Negative	Charging	Curtailed	No
5	Islanded	Night	Low/High	Positive	Discharging	—	—
6		Day	Low/High	Positive	Discharging	All in	—
7				Negative	Charging	Curtailed	—

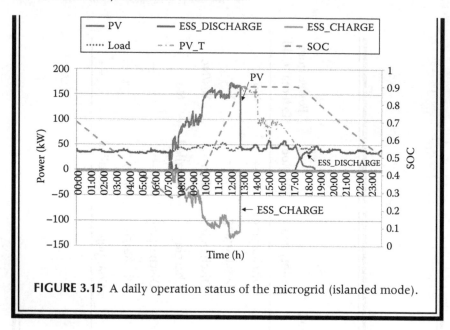

FIGURE 3.15 A daily operation status of the microgrid (islanded mode).

3.4 CELLs

3.4.1 CELL Concept

The idea of a 'CELL' was proposed to tackle the challenges of increasing penetration of distributed generation, especially renewable generation with its variability and uncontrollability. A CELL is used to define a wider area of network in which a collection of DERs can be controlled in response to a number of objectives and can be considered as an entity that consists of a number of microgrids. Its electrical capabilities are defined as a black box and it lends itself to a hierarchical structure in which CELLs can be aggregated or subdivided. From this perspective, a CELL can also be seen as a utility-scale microgrid. The goal of CELLs is twofold:

1. In normal operation, CELLs manage the network optimally, including the DERs and active demands as a whole, and achieve more effective use of the distribution network assets.

2. In a stressed contingency situation, CELLs provide additional possibilities of network operation, such as emergency transfer of the CELL area into islanded operation with preservation of power supply to as many consumers as possible.

3.4.2 CELL Structure and Control

An example CELL structure is shown in Figure 3.16, in which the radially operated 33 kV network below the 132/33 kV transformer is considered as a CELL [5].

To realize control of a CELL, the functions that are required for the necessary data communication, measurements, monitoring and control of the system will include

1. Online monitoring of the total load and generation within the CELL

2. Active power control of the distributed generators

3. Reactive power control of the capacitor banks and/or controllable reactive power from distributed generators

4. Voltage control by utilizing automatic voltage control (AVC) in substations and/or activating automatic voltage regulators (AVR) on some distributed generators

5. Frequency control by activating speed governing systems (SGS) on some distributed generators

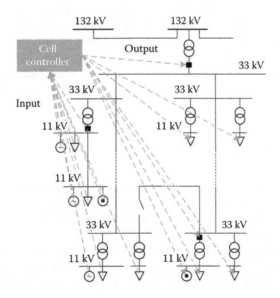

FIGURE 3.16 CELL structure.

6. Capability of remote operation of circuit breakers, reclosers, sectionalizers in substations and feeders and those for the connection of the distributed generators

7. Automatic fast generation or load shedding in case of power imbalance

8. Automatic fast islanding of an entire CELL in case of severe grid fault

9. Voltage, frequency and power control for an islanded CELL

10. Synchronizing the CELL back to a grid-connected system

11. Black-start support to the transmission grid in case of a blackout

Each CELL is required to operate in parallel with the upstream high-voltage-level power system in both normal and stressed contingency conditions. Any fault on the upstream grid will still be handled by the ordinary protection systems such as distance relays on the transmission lines, etc. This is to ensure that during contingencies the power system does not lose power production, short-circuit power, reactive power or spinning inertia by unintentional islanding in the area of the distribution network. The only exception is that during a severely stressed situation, such as in an impending voltage collapse, the CELL controller switches the CELL to islanded operation.

The CELL controller needs the ability to communicate to/from the DNO and the TSO supervisory control and data acquisition (SCADA) systems. It is from the TSO that an on-line signal of an impending voltage collapse is issued from a phasor measurement unit (PMU)-based early warning system. It is also from either the DNO or the TSO that the request to provide black-start support will be sent to the CELL controller.

3.4.3 CELL Perspectives

Efforts have been made to devise a design of a CELL controller that is general enough to allow for inclusion of microgrids to be combined into larger CELLs and even high-voltage TSO grid areas. A network consisting of several CELLs is called a multi-CELLs network. This is illustrated in Figure 3.17, where the general layout of a multi-CELLs network is based on a multi-level control hierarchy and three levels are included in the system. In Level 1, a microgrid is formed by

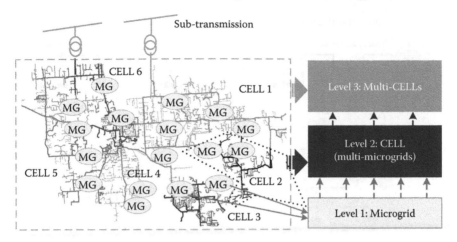

FIGURE 3.17 Perspectives of multi-CELLs.

low-voltage, distributed generation systems, energy storage systems, loads, monitoring and protection devices, etc. In Level 2, a network of several microgrids form a CELL, and multi-CELLs are considered to be the highest level, i.e. Level 3.

The general design is being developed to ensure that network companies can control multiple CELLs by adding a Level 3 controller which could be embedded in the distribution company's SCADA system. Furthermore, it is envisioned that even the national TSO can gain access to all CELLs in the entire power system by adding a controller which in turn would be embedded in the SCADA system of the TSO.

3.5 Virtual Power Plants

3.5.1 VPP Concept

A VPP is used to characterize a group of DERs, and it is comparable to a conventional power plant connected to the transmission network, as shown in Figure 3.18.

A transmission-connected power plant is able to make contracts to sell its electricity and interact directly with other market participants to offer services for power system operation through interactions in the wholesale market or direct contact with transmission system operators, energy suppliers and other parties.

Actual network configuration   VPP perspective

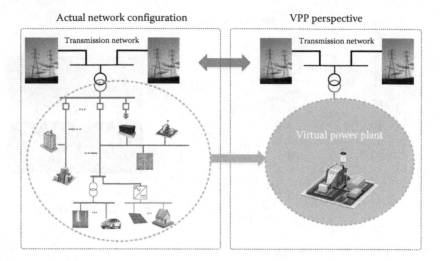

FIGURE 3.18 Characterization of DERs as a VPP.

On its own, a small DER usually does not have access to the electricity markets due to its small capacity, limited flexibility and controllability.

A VPP is a flexible representation of a portfolio of DERs. A VPP not only aggregates the capacity of many diverse DERs, but also creates a single operating profile from a composite of the parameters characterizing each DER and also incorporates network constraints into its description.

A VPP is characterized by parameters usually associated with a traditional transmission-connected generator, such as scheduled output, ramp rates, voltage regulation capability, reserve, etc. A VPP incorporates controllable demand and so parameters such as demand price elasticity and load recovery patterns will be also used to characterize the VPP. Table 3.2 gives an example of generator and controllable load parameters that can be aggregated and used to characterize a VPP.

A VPP is able to facilitate trading by DERs in various energy markets (e.g. forward markets and the power exchange) and provide services to support transmission system operation (e.g. various types of reserve, frequency and voltage regulation). A VPP can also contribute to the active management of distribution systems. The activities of market participation and system operation support of a VPP are categorized as 'commercial' and 'technical' activities, which gives two potential roles for a VPP: commercial VPP (CVPP) and technical VPP (TVPP).

A CVPP is a competitive market actor (e.g. an energy supplier). The composition of a CVPP portfolio is not constrained by location, although for participation in some markets the DER may have to be managed within a geographical location. The DER will contract with a CVPP to

TABLE 3.2 Examples of Generation and Controllable Load Parameters for Aggregation to Characterize a VPP

Generator Parameters	Controllable Load Parameters
• Schedule or profile of generation • Generation limits • Minimum stable generation output • Firm capacity and maximum capacity • Stand-by capacity • Active and reactive power loading capability • Ramp rate • Frequency response characteristic • Voltage regulating capability • Fault levels • Fault ride-through characteristics • Fuel characteristics • Efficiency • Operating cost characteristics	• Schedule or profile of load • Elasticity of load to energy prices • Minimum and maximum load that can be rescheduled • Load recovery pattern

optimize its revenue and visibility in the energy and system management markets. The CVPP will manage the DER portfolio to make optimal decisions on participation in these markets. A single distribution network area may contain more than one CVPP aggregating DERs in its region. A DER is free to choose which CVPP offers most favourable representation/compensation.

Because of the locational requirements in the provision of system management services, a TVPP is usually a monopoly role undertaken by the distribution system operator (DSO) as they are the only party with access to the necessary local system information. Therefore, a TVPP will include every DER in a distribution network region and will be able to present an accurate picture of the network at the point of connection with the transmission system, as well as calculate the contribution of each DER taking location and network constraints into consideration.

A TVPP will use DER operating and cost parameters (received via CVPPs) and local network knowledge to manage the local system and to calculate the characteristics of the network at the point of connection of the distribution and transmission systems for submission to the transmission system operator to assist with transmission level balancing.

3.5.2 VPP Management

3.5.2.1 Commercial VPP

A CVPP provides the following:

- Visibility of DERs in the energy markets
- DER participation in the energy markets
- Maximization of value from DER participation in the markets

A CVPP reduces imbalance risk, and provides benefits of diversity of resource and increased capacity achieved through aggregation. DERs can then benefit from economies of scale in market participation and so maximize revenue opportunities.

In systems allowing unrestricted access to wholesale markets (i.e. any system constraints caused by contracts in the market, or other locational issues are not accounted for at the time of contract creation), CVPPs can represent DERs from any geographic location in the system. However, in markets where energy resource location is critical, the CVPP portfolio will be restricted to include only DERs from the same location (e.g. a distribution network area or a transmission network node). In these instances, a CVPP can still represent DERs from various locations, but the aggregation of resources must occur by location, resulting in a set of DER portfolios defined by geographic location. This is required, e.g., in markets based on locational marginal pricing (transmission system) and in markets where a zonal approach is taken to participation.

As part of optimizing the performance of its DER portfolio, when activity in the wholesale markets is complete, the CVPP will submit to the TVPP information on individual DER contract schedules and corresponding bids and offers (marginal cost) to adjust that position.

Figure 3.19 summarizes CVPP activities. Each DER included in the CVPP portfolio submits information on its operating parameters, marginal cost characteristics, etc. These inputs are aggregated to create a single VPP profile representing the combined capacity of all DERs in the portfolio. With the addition of market intelligence, the CVPP will optimize the revenue potential of the portfolio making contracts in the electricity markets and submit information on the DER schedule and operating costs to the TVPP.

This commercial VPP role can be undertaken by a number of market actors, including incumbent energy suppliers, third-party independents or new market entrants. DERs are free to choose a CVPP to represent them in the wholesale market.

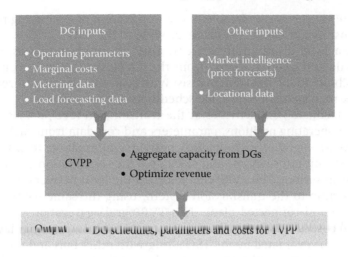

FIGURE 3.19 Inputs to and output from CVPP activities.

3.5.2.2 Technical VPP

A TVPP provides the following:

- Visibility of DERs to the system operator(s)

- DER contribution to system management

- Optimal use of DER capacity, providing system balancing at the lowest cost

The technical VPP aggregates and models the response characteristics of a system containing DERs and networks within a single geographical area. A hierarchy of TVPPs may be created to systematically characterize the operation of DERs at low-, medium- and high-voltage regions of a local network, but at the transmission–distribution network interface, the TVPP presents a single profile representing the entire local network. This technical characterization is equivalent to the characterization that the transmission system operator has of transmission-connected generation and corresponding transmission network topology.

A TVPP acquires information on each DER within its local network region to facilitate the active management of the local network and the technical characterization/grid aggregation of the network at the transmission level. This information can be submitted by the CVPP on behalf of the DERs. This is comparable to the notification of generating position that transmission-connected plants provide to the TSO.

In a local distribution network, DER operating positions, parameters, bids and offers collected from the CVPP can be used to improve DER visibility to the DSO and to assist with real-time or close-to-real-time network management to provide scheduled ancillary services.

To facilitate DER activities at the transmission level, a TVPP aggregates the operating positions, parameters and cost data from each DER in the network together with detailed network information (e.g. topology and constraint information) and calculates the contribution of each DER in the local system. The TVPP characterizes the local network at its point of connection to the transmission system, using the same parameters as a transmission-connected plant. This TVPP grid aggregation profile and marginal cost calculation (reflecting the capabilities of the entire local network) can be evaluated by the TSO along with other bids and offers from transmission-connected plants to provide real-time system balancing.

Carrying out TVPP activities requires local network knowledge and network control capabilities. The DSO usually is best placed to fulfil this role. With this TVPP capability, the DSO role can evolve to include the active management of the distribution network, analogous to a transmission system operator. The DSO will continue to be a local monopoly.

Figure 3.20 summarizes the TVPP activities. TVPP activities involve the management of local network constraints (i.e. active distribution network management) and grid aggregation of a distribution network area to characterize the contribution and characteristics of the local network at

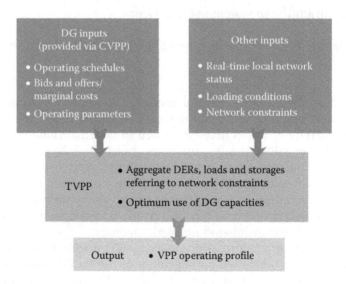

FIGURE 3.20 Inputs to and outputs from TVPP activities.

its point of connection with the transmission system. TVPPs are composed of all the DERs in a single distribution network area.

Information on DERs in the local network is passed through to the TVPP by the various CVPPs that represent DERs in the area. The TVPP will use this information in conjunction with detailed network information, e.g. topology and network constraints, to characterize the contribution of the distribution network (and associated generation and loads) at the point of connection to the transmission system. This characterization can be used at the transmission level to provide transmission system balancing services.

3.5.2.3 *Interactions between CVPP and TVPP*

Figure 3.21 illustrates the respective roles of the CVPPs and TVPPs and their interactions with each other and the wider markets. The CVPP is operational in the energy markets and is responsible for passing

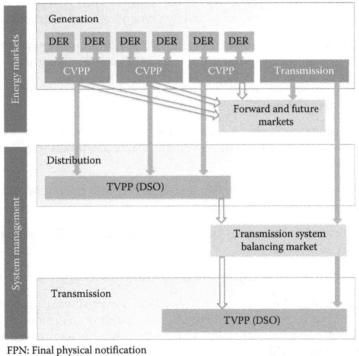

FPN: Final physical notification

> ⟹ Individual operating parameters, contracts, or FPN/bids and offers
> ⟹ Aggregated operating parameters, contracts, or FPN/bids and offers

FIGURE 3.21 CVPP and TVPP activities in the energy market and system management context.

information on DER through to the TVPP. The TVPP is engaged in system management and facilitates management of local network constraints as well as aggregating DERs with local network parameters for presentation at the transmission level.

The CVPP optimizes the position of its portfolio with reference to wholesale markets and passes DER schedules and operating parameters on to the TVPP. The DSO/TVPP uses input from the CVPPs operating in its area to manage any local network constraints and determines the characteristics of the entire local network at the transmission connection point. This can then be used to offer transmission balancing services which the transmission system operator can evaluate along with bids and offers from transmission-connected generation.

☐ Questions

1. List at least four services provided by DSI and discuss their differences.

2. Explain how DSI provides frequency support for a power system, and justify its superiority to traditional partially loaded generators.

3. Judge whether the following statements are correct or not. If not correct, give the reasons.

    a. The power sources of a microgrid are renewables such as wind power, solar power, tidal power, etc. There are no diesel generators in a microgrid.

    b. Energy storage systems play a crucial role in a microgrid and must be present in every microgrid.

    c. A microgrid always provides power with high reliability, which is its main advantage.

    d. There are two types of microgrids: AC microgrids and DC microgrids.

    e. The microgrid is a good solution to the power shortage problem of islands or remote areas. Microgrids on islands or in remote areas are all independent microgrids, that is they are not connected to the main power grid.

4. What are the ambitions of CELLs? List the key functionalities for common CELL controllers.

5. What are the differences between CVPP and TVPP? How do they interact with each other?

☐ References

1. J. Manwell, A. Rogers, G. Hayman et al., *Hybrid2 – A Hybrid System Simulation Model: Theory Manual*. Boston, MA: University of Massachusetts, 2006.
2. L. Guo, W. Liu, X. Li et al., Energy management system for stand-alone wind-powered-desalination microgrid, *IEEE Transactions on Smart Grid*, 7(2), 1079–1087, 2016.
3. B. Zhao, X. Zhang, P. Li et al., Optimal sizing, operating strategy and operational experience of a stand-alone microgrid on Dongfushan Island, *Applied Energy*, 113, 1656–1666, 2014.
4. C. Wang, B. Jiao, L. Guo et al., Robust scheduling of building energy system under uncertainty, *Applied Energy*, 167, 366–376, 2016.
5. Energinet.dk, Cell Controller Pilot Project, Energinet.dk, Fredericia, Denmark, 2011.

CHAPTER 4

ICT Infrastructure and Cyber-Security

4.1 ICT Infrastructure and Cyber-Security

Smart electricity distribution networks utilize advanced technologies that integrate automated control and communications infrastructure from the grid substation to the smart meters at customer premises. They enable efficient, reliable and safe operation of the distribution networks while facilitating a smooth integration of renewable energy resources. The intelligent monitoring and control enabled by modern information and communication technologies (ICT) have become essential to realize the following objectives envisioned from smart electricity distribution networks [1]:

- Minimizing the power disturbances and outages due to equipment failures, capacity constraints and natural accidents and catastrophes through online power system condition monitoring, diagnostics and protection systems

- Providing real-time predictive information and corresponding recommendations to all stakeholders (e.g. utilities, suppliers and consumers) so as to optimize their power utilization

- Offering services such as intelligent appliance control for energy efficiency and better integration of distributed energy resources (DERs)

- Offering automatic control actions that ensure the quality, security
  and reliability of the power supply

4.2 Communications Technologies

With the emerging smarter electricity distribution networks that integrate advanced technologies and applications, a large amount of data
needs to be transmitted between the smart network entities and the electric utilities for analysis, control and real-time pricing. In order to provide
a reliable, secure and cost-effective service, it is crucial for the electric
utilities to define the communication requirements and find the best
communications infrastructure to transmit data within their networks.

Different communication technologies supported by wired and
wireless media are available for data transmission. Even though wireless
communications have some advantages over wired technologies, such as
low-cost infrastructure and ease of connection to difficult or physically
unreachable areas, the nature of the transmission path may attenuate or
distort the signal. On the other hand, a properly designed and installed
wired solution does not have interference problems and its functions
are not dependent on batteries, as wireless solutions often do. For communications between various field sensors and equipment and supervisory control and data acquisition (SCADA) systems and smart meters,
power line communication (PLC) or wireless communications are used.
For communications from smart meters to utility's data centres, cellular
communication networks and digital subscriber lines are employed.

4.2.1 Wireless Mesh

A mesh network is a flexible network consisting of a group of nodes,
where new nodes can join the group and each node can act as an independent router. This self-healing nature of the network enables the communication signals to find another route via the active nodes, if any node
drops out of the network. Hence, it enhances reliability.

In a wireless mesh network, each node, in addition to sending its
own data, acts as a signal repeater until the collected data reach the electric network access point. Then, the data are transferred to the utility via
a communication network, as shown in Figure 4.1.

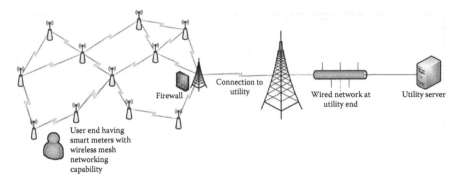

Firewall

Connection to utility

Wired network at utility end

Utility server

User end having smart meters with wireless mesh networking capability

FIGURE 4.1 Typical mesh network connecting remote smart meter mesh network to the utility network through wired and wireless infrastructure.

The dynamic self-organization, self-healing and self-configuration features offer improved network performance, balancing the load on the communication network, extending the network coverage range in urban and suburban areas with the ability of multi-hop routing.

Limited network capacity, slow and fast fading and susceptibility to interference are some of the main disadvantages of wireless mesh networking systems. In urban areas, mesh networks face a coverage challenge since the node density cannot provide complete coverage of the communications network. Coverage in urban areas is at risk from non-uniform node density dispersion as this is largely decided by the self-healing feature of the nodes. Furthermore, a third-party company is required to manage the network. Since the metering information passes through every access point, encryption techniques are applied to the data for security purposes.

ZigBee and ZigBee Smart Energy Profile (SEP) have been recognised as the most suitable wireless mesh communication specifications for smart grid residential network domains by the National Institute for Standards and Technology (NIST) of the United States [2,3]. ZigBee SEP supports utilities in sending messages to home owners, who can also obtain information of their real-time energy consumption.

ZigBee is a wireless communication technology with relatively low power consumption, data rate, complexity, and cost of deployment. ZigBee specification is based on the IEEE 802.15.4 standard. ZigBee has 16 channels in the 2.4 GHz band, each with 5 MHz of bandwidth. Each channel is 2 MHz wide and has a 3 MHz guard band. The maximum output power of ZigBee radios is 0 dBm (1 mW), with a transmission range between 1 and 100 m and a 250 kbps data rate and OQPSK (offset quadrature phase shift keying) modulation.

Example 4.1

It is required to assign 16 channels in ZigBee to a frequency spectrum having a lowermost frequency of 2402 MHz. Calculate the uppermost frequency of the frequency spectrum.

Answer

In ZigBee, each channel needs 5 MHz bandwidth; hence the bandwidth required for 16 channels is (5 × 16 MHz) 80 MHz. Since the lowermost frequency is 2402 MHz, the uppermost frequency is (2402 + 80 MHz) 2482 MHz.

Thanks to its simplicity, mobility, operation within an unlicensed spectrum and easy network implementation, ZigBee is an ideal technology for smart lighting, energy monitoring, home automation and automatic meter reading.

The main demerits of ZigBee in general are low processing capabilities, small memory size, small delay requirements between consecutive incoming messages and being subjected to interference with other appliances, which share the same transmission medium. Currently it shares the industrial, scientific and medical (ISM) frequency band with IEEE 802.11 wireless local area networks (WLANs), WiFi, Bluetooth and microwave. This lack of robustness increases the possibility of corrupting the entire communications channel due to the interference of IEEE 802.11/b/g in the vicinity of ZigBee.

4.2.2 Cellular Network Communication

In some smart distribution network applications, a large amount of data will be generated and a connection with a high data rate would be required. Existing cellular networks can be a good option for communicating between smart meters and utilities. This approach avoids operational costs and additional time for building a dedicated communications infrastructure.

Global system mobile (GSM) technology having data rates up to 14.4 kbps and global packet radio system (GPRS) having data rates up to 170 kbps are employed for cellular networks. These technologies can support advanced metering infrastructure (AMI), Demand Response (DR) and home area network (HAN) applications. Lower cost, better coverage, lower maintenance costs and fast installation features highlight why cellular networks can be the best candidate as a communications technology for such applications. Cellular networks have the capacity to deliver large amounts of data in a short period of time. They also offer secure transmission of data.

For hard real-time applications, continuous availability of communication channel is mandatory. However, the services of cellular networks are shared by customer markets and their reliability is low under abnormal situations, such as a windstorms. Therefore, for time-critical applications, utilities rely on their own private communications network.

4.2.3 Power Line Communication

PLC is a technique that uses existing power lines to transmit high-speed (2–3 Mbps) data signals from one device to another.

PLC has been the first choice for communication with electricity meters due to its capacity to directly connect with the meters and the successful implementations of AMI in urban areas where other solutions struggle to meet the needs of utilities.

In a typical PLC network, smart meters are connected to the data concentrator through power lines and data are transferred to the data centre via cellular network technologies, as shown in Figure 4.2.

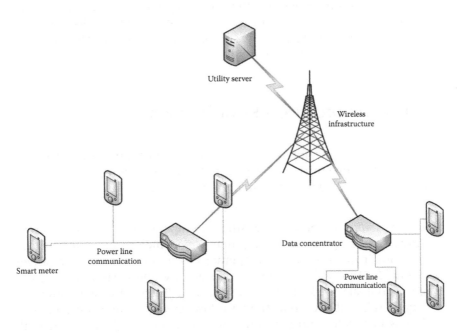

FIGURE 4.2 PLC and wireless technology hybrid.

The use of existing electrical infrastructure decreases the installation cost of the communications infrastructure, hence making PLC an attractive communications technology.

Unlike in peer-to-peer connections, data transmissions in PLC are broadcast in nature, hence the data is less secure.

The low bandwidth (20 kbps) restricts the applications of PLC. The network topology, number and type of devices connected to the power lines and wiring distance between transmitter and receiver adversely affect the quality of signals transmitted over the power lines. In addition, the communication becomes sensitive to disturbances on the power line, which makes the quality of service of the communication in PLC dependent on the quality of the power signals.

4.2.4 Digital Subscriber Lines

Digital subscriber lines (DSLs) is a high-speed digital data transmission technology that uses the wires of the voice telephone network. The already existing infrastructure of DSL lines reduces the installation cost. The widespread availability, low cost and high-bandwidth data transmissions make the DSL technology an ideal candidate for smart distribution network applications. However, the reliability and potential down time of the DSL technology may not be acceptable for time-critical applications. Furthermore, wired DSL-based communications systems require cables to be installed and regularly maintained.

4.3 Communications Requirements

The different requirements of the communications infrastructure between different entities of a smart distribution network should be determined by considering the following attributes:

1. *Interoperability*: In order to ensure interoperability among different plants, sensors, protection equipment, smart meters, etc., the communications infrastructure may operate in the lower frequency ranges (below 2 GHz). This ensures a high-quality and cost-effective communication across the utility's service area. The operating frequencies may be selected such that they are low enough to overcome line-of-sight issues such as rain fade and penetration through walls and high enough to overcome undesirable impacts of noise.

2. *Data rate*: This defines how fast the data is transmitted between the smart distribution network components. The data rate requirements can be different for each specific application. Some of the applications that support system emergencies require high data rates to achieve a reliable and accurate data transfer. On the other hand, the communication data rates for distribution automation and AMI can be low. The total throughput* that the smart grid will require for the communications systems with many applications, for example AMI for demand response, is estimated at 3–10 Mbps.

3. *Reliability*: It is important to check[†] how reliably a communications system can perform data transfer according to the specific requirements. The communication nodes should always be reliable for the continuity of communications. Some smart distribution network applications, such as distributed automation, expect highly reliable data communication, and some can tolerate outages in data transfer.

4. *Latency*: This is the delay of the data transmitted between the different entities of smart distribution networks. Some mission-critical, real-time applications such as distribution automation systems, may not tolerate any latency. For other soft, real-time applications, such as AMI or home energy management (HEMs), latency is not critical.

4.3.1 Communication Requirements of Substation Automation

Substations are key elements of the power grid network and all their devices are monitored, controlled and protected by substation automation (SA). The SA system collects data and performs on-site protection and control while sharing data with the control centres; this ensures the robust routing of power from generators to loads through the complex network of transmission lines. The communication network plays a critical role within SA to enable full control and monitoring of real-time operating conditions and performances of substations. Figure 4.3 shows a star-connected SA system. Generally an SA system comprises three levels: (1) the station level, which includes the substation computer, the substation human–machine interface and the gateway to the control

---

* Throughput is the rate at which data reaches the destination. Since the data undergoes delays and other disturbances, throughput is always less than data rate.
† Reliability is checked by implementing various handshaking techniques, bit error checking and bit error correction techniques.

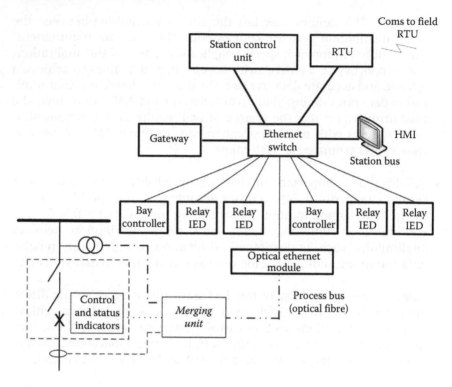

FIGURE 4.3 Star-connected SA system.

centre; (2) the bay level, which includes all the controllers and intelligent electronic devices (IEDs); and (3) the process level, which consists of switchgear control and monitoring, current transformers (CTs), voltage transformers (VTs) and other sensors. These levels are connected by two communication buses: the station bus and the process bus. Even though in Figure 4.3 the station bus is connected in star, depending on the design it could also be ring connected or a combination of both.

The *merging unit* combines the signals received from conventional and optical current and voltage transformers, and the combined digital signals are sent to multiple receiving units, such as IEDs, displays and station computers. Fibre optics are the preferred communication media for the process bus as the wired technologies need high protection from the hazardous electrical environment within the substation bay.

For the station bus, a highly reliable, scalable, secure and cost-effective communications network is a prerequisite. The latency requirements must be low, sometimes as low as a few power frequency cycles, to prevent communication from timing out.

Example 4.2

A *merging unit* transmits a data set containing three-phase current and neutral current and three-phase voltage and neutral voltage through an optical fibre to the station bus.

a. An optical fibre is used for the process bus with an SA system. The fibre is a step-index multimode fibre having a core of refractive index 1.5 and cladding of refractive index 1.485. If the length of the fibre is 200 m, what is the maximum data rate possible?

b. According to the IEC 61850-9-2LE standard, the recommended sampling rate of each signal is 80 samples per period. Each sample needs 32 bits. Is the data rate offered by the fibre optics sufficient?

Answer

a. Modal dispersions (how much a light signal is spread in time because the propagation velocity of the optical signal is not the same for all modes) limit the data rate within an optical fibre. Data rate into transmission length (BL) should be less than $(n_2/n_1^2) \times (c/\Delta)$ for a step indexed fibre, where $\Delta = (n_1 - n_2)/n_1$ and c is the speed of the light [4].

$$\Delta = \frac{1.5 - 1.485}{1.5} = 0.01$$

$$B \times 200 < \frac{1.485}{1.5^2} \times \frac{3 \times 10^8}{0.01} = 1.98 \times 10^{10}\,\text{bits/s}$$

The maximum data rate possible is $0.99 \times 10^8\,\text{bits/s} = 99$ Mbps.

b. Since the period of the 50 Hz signals is 20 ms, the sampling rate $= 80/(20 \times 10^{-3}) = 4000$ samples/s.

Each data set contains 8 samples, and since each sample occupies 32 bits of data, the data rate required $= 4000 \times 8 \times 32$ bit/s $= 1.02$ Mbps.

The optical fibre can transmit the data set at the required rate.

4.3.2 Communication Requirements of Overhead Distribution Line Monitoring

Overhead lines are vulnerable to overheating, lightning strikes and icing. Hence, to monitor overhead lines, wireless sensor nodes are deployed and they communicate with the IEDs continuously.

The communication requirements for overhead line monitoring systems depend on the network model, number of nodes and the preferred communication technology. Importantly, a large portion of energy flows through transmission lines; hence, overhead transmission line monitoring systems should support reliable, secure, effective and real-time communication to respond to emergency situations quickly.

4.3.3 Communication Requirements for Advanced Metering Infrastructure

AMI supports both consumers and utilities. It provides consumers energy consumption data, pricing information, etc. and allows utilities to collect and analyse energy consumption data of all consumers connected to their network. Figure 4.4 outlines the typical components in the AMI.

A home area network (HAN) focuses on power management on the consumer side, where home appliances are monitored and controlled to balance and optimize power supply and consumption. A HAN basically consists of smart meters, smart appliances, in-home displays, and advanced control systems. If a HAN is used to send measured energy consumption data to an in-home display, its communication needs can be handled with low-power, short-distance technologies, such as ZigBee and Bluetooth. There is no need for a large bandwidth or high communication speed. However, if the HAN is used to communicate with smart appliances for direct load control or demand response, then the communication requirement will scale up.

As shown in Figure 4.4, the measured data from homes will be sent to a concentrator through a neighbourhood area network (NAN). The communication requirement of a NAN depends on the number of homes connected and the functional requirement. For example, if functions like direct load control or automatic control actions through real-time pricing are required, then a connection to each smart appliance within a home is essential, and thus the communications requirements will be moderately high. Many AMI deployments that are planned or are in operation use wireless mesh networks for the NAN. For example, the AMI infrastructure in Gothenburg city uses ZigBee for the NAN.

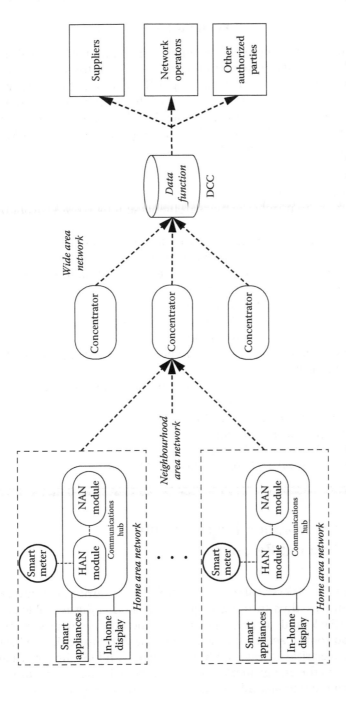

FIGURE 4.4 Typical components of the AMI. (From Samarakoon, K.B., Use of smart meters for frequency and voltage control, PhD thesis, School of Engineering, Cardiff University, Cardiff, UK, 2012.)

Example 4.3

A communication network in which a smart meter installed in each house sends measurements to a concentrator installed at the 11/0.4 kV transformer, forming a NAN, is shown in Figure 4.4. There are 400 smart meters connected to the concentrator. Assuming two bytes are used to send each measurement from the smart meter, calculate the data rate required within the NAN when the sampling interval is 15 and 30 s.

Answer

When the sampling interval is 15 s:
The maximum number of measurements sent by HEMS through a NAN per min
  = No. of measurements per min × number of smart meters
  = (60/15) × 400 = 1600 measurements
As each measurement needs 2 bytes to transmit data, the number of bits transferred per min
  = 1,600 × 2 × 8 = 25,600 bits/min
Data rate required = 25,600/60 bits/s = 427 bps
When the sampling interval is 30 s:
The maximum number of measurements sent by HEMS through a NAN per min
  = No. of measurements per min × number of smart meters
  = 2 × 400 = 800 measurements
As each measurement needs 2 bytes to transmit data, the number of bits transferred per min
  = 800 × 2 × 8 = 12,800 bits/min
Data rate required = 12,800/60 bits/s = 213 bps

4.3.4 Communication Requirements of Demand Response

Demand response (DR) entails (1) the control of loads during peak and valley loading conditions, thus ensuring a better utilization of the available energy and more reliable and cheaper operation of the entire power system; (2) dynamic load control actions during loss of generation to achieve a balance between the electrical energy supply and demand; and (3) the participation of consumers in the energy markets by changing their energy consumption, thus allowing them to take advantages of different pricing regimes.

The communication requirements of a DR application depend on its purpose. If DR is used to send a shut-off command or a new operating set point to home appliances, such as heaters and air conditioners, then communications requirements are low, thus requiring low bandwidth. If DR is used for a mission-critical application such as load-balancing during a system emergency, a lower latency is a must.

4.3.5 Communication Requirements of Outage Management Systems

Power outages are inevitable in power systems due to short circuits in plants or lines and other failures at power stations. Power outages result in financial losses to the utilities and customers. Hence, outage detection, management and restoration are very critical to provide reliable and quality electricity supply.

An outage management system (OMS) is a system with a trouble call centre, computer-based tools, and utility procedures to efficiently and effectively identify faults, diagnose and locate fault, provide feedback to affected customers, dispatch trouble/repair crews, restore supply, maintain historical records of the outage and calculate statistical indices on electrical outages. In OMS, utilities with very limited penetration of real-time systems but good customer and network records use a trouble calls to detect a fault, whereas those with good real-time systems and extended control can use direct measurements from automated devices. For example, the last gasp messages from smart meters can be used as an input to the OMS. Fault diagnosis and fault location algorithms will operate more efficiently and effectively with such additional data points.

The automated restoration of an outage in a system highly depends on fault identification, diagnostics and location phases. The communication requirements for the quick restoration of an outage are high, and, typically, a latency less than 2 s and a bandwidth of 56 kbps are preferred [6,7].

4.3.6 Communication Requirements of Distribution Automation

Passive distribution networks that simply routed electric power from the transmission system to end-users are now becoming more active thanks to distribution automation (DA). A number of different DA solutions are emerging to automatically and remotely monitor, control and coordinate distribution components in real time. These include voltage control through capacitor bank switching and improved fault detection, isolation and restoration. Most DA applications are time critical and therefore the latency of the communication system should be minimum. Latency less than 1 s for alarms and less than 100 ms for control have been reported [6,7].

4.3.7 Communication Requirements of Distribution Management System

Distribution networks have become too complex to be controlled and monitored manually. An advanced distribution management system (DMS) is required to visualize the distribution networks and monitor any abnormalities in the system. The DMS manages, operates and maintains the power distribution assets and ensures reliable, secure and efficient power delivery.

The DMS has a suite of applications that are closely integrated with real-time monitoring and control systems and provides complete management and maintenance functions for the distribution network. Thus, highly reliable and seamless communication with real-time systems, a strong integration capability and advanced interoperability between other components are the most crucial requirements for DMS.

4.3.8 Communication Requirements of Asset Management

Electric utilities have been under pressure to assure a QoS (quality of service) at minimum cost. Asset management has primarily been developed as a response to this problem by offering management, automation, tracking, and optimization of work order processes, field crews scheduling and field assets.

Equipment condition monitoring, coordinated asset management and dynamic adjustment of operating limits are the critical functions of an asset management system. Thus, advanced monitoring devices and seamless data traffic with other applications, such as SCADA, GIS, AMI, are essential.

4.3.9 Communication Requirements of Distributed Energy Resources

With the increased penetration of renewable energy resources (DER) such as solar and wind power, electric vehicles, combined heat and power and energy storage, maintaining the voltages in the distribution network within the statutory limits and maintaining the balance between the generation and demand in real time become a challenge. To overcome these challenges, the following are considered:

1. Smoothing the output power variation of renewable generation plants, such as wind and solar, by coordinated operation with energy storage or batteries in a fleet of vehicles
2. DR initiatives to shift or control loads to follow generation
3. Coordinated control of DERs with on-load tap changers and capacitor banks

These operations require a fast and a reliable communications network that can monitor and control DERs in real time.

Further, a microgrid containing many DERs is now emerging. Even though islanded operation of a microgrid is not allowed by many utilities at present, such an operation requires fast communication among local generation and loads in order to maintain the stability of the microgrid.

4.3.10 Communication Requirements of Grid to Vehicle and Vehicle to Grid

Electric vehicles (EVs) absorb power to charge their batteries (grid to vehicle [G2V]) and the stored energy is then available to support the grid if required (vehicle to grid [V2G]).

One of the concerns about G2V operation is if the charging is not properly scheduled, EV charging could cause peak demand to increase drastically. Therefore, coordinated charging is a must. Ideally, if EV charging can be shifted to off-peak periods, that will not only minimize the increase in peak demand but also support the power system by valley filling. However, the coordinated operation of the EV demands a good communications infrastructure. The communication requirements differ based on whether a fleet of vehicles are dispersed in homes or concentrated in a multi-car charging facility.

EVs under V2G operation act as distributed resources in that the power previously absorbed from the grid or produced by the kinetic energy during motion can be sent back to the utility. One of the important roles of V2G is the ability to support renewable energy. As discussed in Section 4.3.9, the penetration and the intermittency of renewable energy can be problematic and can be smoothened out by using EVs operating in the V2G mode. The communication needs of V2G should be considered for a fleet of vehicles and dispersed vehicles.

The communication needs of a fleet parked in one location are simple. A short-range, lower cost wireless communication technique, such as ZigBee or Bluetooth, can be used for each parking space. Long-distance communication facilities are needed for dispersed vehicles for electronic identification of the electric utility meter that the vehicle is plugged into, for billing systems and for capacity identification. Hence, cellular communication technologies and land-line communication are the best choices for these purposes.

Based on a number of consultations, the U.S. Department of Energy has come up with a summary of communication requirements for smart distribution networks; these are summarized in Table 4.1.

TABLE 4.1 Communication Requirements for Different Smart Distribution Network Functions

Application	Bandwidth (kbps)	Latency	Security
AMI	10–100	<2000 ms	High
DR	14–100	<500 ms	High
OMS	56	<2000 ms	High
DA	9.6–56	20–200 ms	High
Asset management	56	<2000 ms	High
DMS	9.6–100	100–2000 ms	High
DER	9.6–56	300–2000 ms	High
G2V and V2G	9.6–56	2000 ms–5 min	High

Sources: Gungor, V.C. et al., *IEEE Trans. Ind. Inform.*, 9(1), 28, February 2013; Communications requirements of smart grid technologies, Department of Energy, Washington, DC, 2010.

4.4 Distribution Communication Standards

There are many applications, techniques and technological solutions for smart electricity distribution systems that have been developed or are still in the development phase. The American National Standards Institute (ANSI), the International Electrotechnical Commission (IEC), the Institute of Electrical and Electronics Engineers (IEEE), the International Organization for Standardization (ISO), the International Telecommunication Union (ITU), etc. [8] are leading the standardization efforts for seamless interoperability, robust information security, increased safety of new products and systems and a compact set of protocols and communication exchange that are required for smart distribution networks. Table 4.2 summarizes the different standards available for remote metering, SA, power line networking, and inter-control and interoperability centre communications. Furthermore, some of the frequently used standards are described in subsequent sections.

4.4.1 IEC 61850

IEC 61850 is a collection of protocols and a comprehensive standard that provides protocols for data transmission, definition of data, services and behaviour, standardized object names, communications services that support a variety of services and standardized substation configuration language (SCL). It requires an Ethernet physical network and typically uses high-speed switches to provide network connectivity. With Ethernet as the physical layer, other protocols such as Modbus and DNP can co-exist on the same network.

The IEC 61850 network that provides device-to-device communications is termed the 'station bus'. The standard also defines a mechanism for the continuous transfer of digitized process data, such as currents and voltages and distributing these digitized signals to multiple IEDs. This communications network is defined as a process bus. Currents and voltage measurements collected by current and voltage transformers are digitized by field units called merging units and passed to substation equipment through the process bus [9]. This architecture is shown in Figure 4.5.

The data model specified in IEC 61850 starts with a physical device. Each physical device is then divided into a number of logical devices. Each logical device has a number of logical nodes, each representing logically grouped power system functions. More than 90 such logical nodes are explicitly defined in IEC 61850. Then data related to each logical

TABLE 4.2 Different Standards Used in Smart Distribution Networks

Application	Standard	Description
Remote Metering Standards	ANSI C12.19	This is an ANSI standard for utility industry end device data tables. It defines a table structure for data transmissions between an end device and a computer for utility applications using binary codes and XML (Extensible Markup Language) content. This standard does not define device design criteria, device language specifications or the protocol used to transport that data.
	ANSI C12.18	This ANSI standard is specifically designed for meter communications and is responsible for two-way communications between a C12.18 device (e.g. smart electricity meters) and a C12.18 client via an optical port.
	M-Bus	This is an European standard and provides the requirements for remotely reading all kinds of utility meters. The utility meters are connected to a common master that periodically reads the meters via M-Bus. The wireless version, Wireless M-Bus, also exists.
Substation Automation Standards	IEC 61850	IEC 61850 is a standard for the design of the SA and control communication system. IEC 61850 is a part of the IEC Technical Committee 57 (TC57) communication reference architecture for electric power systems.
	IEC 61499	The IEC 61499 Standard is for the development, reuse and deployment of function blocks (FB) in distributed and embedded industrial control and automation systems.

(Continued)

TABLE 4.2 (*Continued*) Different Standards Used in Smart Distribution Networks

Application	Standard	Description
Power Line Networking Standards	PRIME	PRIME is an open, global power line standard that provides multivendor interoperability and welcomes several entities to its body.
	G3-PLC	G3-PLC is a PLC specification launched by the ERDF and Maxim companies that aim to provide interoperability, cyber-security and robustness and reduce infrastructure costs in smart grid implementations worldwide.
Application-Level Energy Management Systems Standards	IEC 61970	This standard provides a common information model (CIM) for exchanging data between networks and devices in the transmission domain.
	IEC 61968	IEC 61968 provides a CIM for exchanging data between networks and devices in the distribution domain.
	OpenADR	OpenADR is a research and standards development effort that is defined as a fully automated demand response using open standard, platform-independent and transparent end-to-end technologies or software systems.

(*Continued*)

TABLE 4.2 (*Continued*) Different Standards Used in Smart Distribution Networks

Application	Standard	Description
Inter-Control and Interoperability Centre Communication Standards	IEEE P2030	IEEE P2030 is a guide for smart grid interoperability of energy technology and information technology operation with the electric power system, and customer-side applications. It enables seamless data transfer in a two-way communications manner for electric generation, reliable power delivery and customer-side applications.
	ANSI C12.22	ANSI C12.22 is defining a protocol for transporting ANSI C12.19 utility industry end device table data over networks to achieve the interoperability among communications modules and smart meters by using advanced encryption standards for electronic data, enabling strong, secure communications, including confidentiality and data integrity.
	ISA100.11a	ISA100.11a is an open standard for wireless systems for industrial automation. It focuses on robustness, security and network management requirements of wireless infrastructure and low-power consuming devices that provides large-scale installations. The ISA100.11a standard is simple to use and deploy and provides multivendor device interoperability.
	ITU-T G.9955 and G.9956	The two standards ITU-T G.9955 and ITU-T G.9956 contain the physical layer specification and the data link layer specification, respectively, for narrowband orthogonal frequency-division multiplexing PLCs transceivers for communications via alternating current and direct current electric power lines over frequencies below 500 kHz.

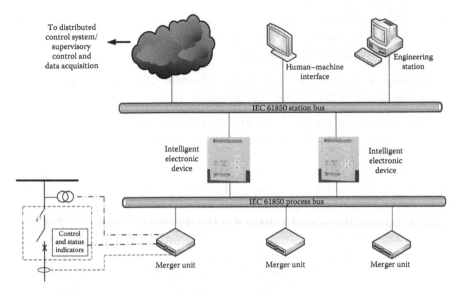

FIGURE 4.5 IEC 61850 system architecture.

node is specified. Typical data elements are system information, device information, measurements, metered values, controllable data, status information and settings. For example, a bay controller is employed for control and monitoring of switchgear, transformers and other bay equipment. The physical device 'bay controller' has a number of logical devices for bay control, protection and metering. Each logical device may have more than one logical node. For example, the logical device 'bay control' has logical nodes such as circuit breaker (CB) control, transformer tap change control, programmable automatic sequence control, etc. This is depicted in Figure 4.6.

IEC 61850 also defines the following communication services for a variety of services within a substation:

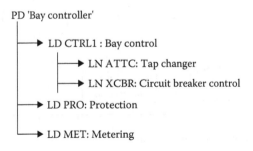

FIGURE 4.6 Logical nodes and devices within the physical device 'bay controller'.

1. For the exchange of binary information over Ethernet, IEC 61850 introduced an information exchange service called generic object oriented substation event (GOOSE).

2. In order to inform any system state changes, IEC 61850 uses generic substation state events (GSSE) messages.

3. The sampled values from CTs and VTs are sent through a serial communication service called sampled values (SV).

For time-critical operations within a substation, information exchange should take place without a delay and within some milliseconds. GOOSE messages are used for such applications. An example with a breaker-failure relay is shown in Figure 4.7. Here the breaker-failure relay monitors the current after a trip signal is issued by the transformer protection. If current is not interrupted then a re-trip signal is sent. If the re-trip signal also fails, then a bus differential protection is triggered. The following are the GOOSE messages that will be communicated during this time-critical event:

1. A transformer protection relay detects a fault and issues a trip signal. At the same time it sends a GOOSE message to the breaker-failure relay.

2. Upon receiving the GOOSE message, the breaker-failure relay waits for a preset period of time, and if the current is not interrupted, it sends a re-trip signal to the circuit breaker.

3. In the event of failure of the re-trip command, the breaker-failure trip sends a GOOSE message to bus differential relay for bus isolation.

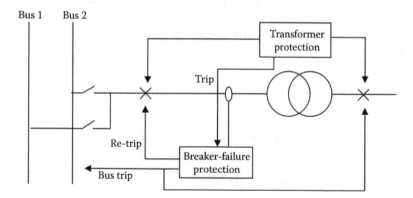

FIGURE 4.7 Breaker-failure protection.

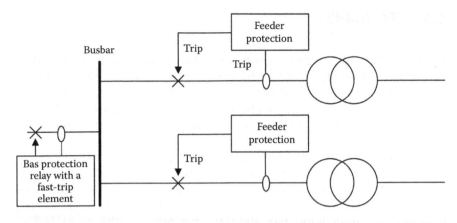

FIGURE 4.8 Fast-trip protection.

In order to inform any system state changes, IEC 61850 uses GSSE messages. For example, in the fast-trip scheme shown in Figure 4.8, when one of the feeder protection relays detects a fault, the following sequence of GSSE messages are communicated:

1. Bus protection relay sends the message 'I see the fault do you see it?' to all the feeder protection relays.

2. The feeder protection relay on the faulted feeder will send the following messages:

   a. I also see the fault

   b. I got your message

3. The fast-trip element of the bus protection is disabled after receiving a message from one of the feeder IED saying 'I also see the fault'.

The sampled values from CTs and VTs are sent through a serial commination service called an SV. As shown in Figure 4.3, the CTs and VTs are connected to a merging unit. This converts the analogue values measured from the CTs and VTs into digital form using an ADC (analogue-to-digital converter) and uses some digital signal processing techniques to send them to the station bus through the process bus as an SV message.

4.4.2 IEC 61499

IEC 61499 Part I defines a function block (FB) as the basic unit for encapsulating and reusing the data and event know-how. This is a software class defining the behaviour of possible multiple instances. As shown in Figure 4.9, FB includes event inputs and event outputs as well as the data inputs and data outputs to provide synchronization between data transfer and program execution in distributed systems.

As the name implies, the basic FB is the 'atom' out of which higher-level 'molecules' are constructed. With the IEC 61499 Part II compliant software tools, software developers can encapsulate FB in the form of algorithms written in one of the IEC 61131 Part III programming languages or other languages such as Java or C++. The execution of these algorithms is triggered by execution control charts (ECCs), which are event-driven state machines similar to the well-known Harel statecharts [10].

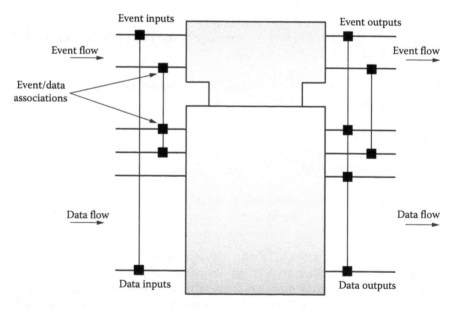

FIGURE 4.9 Function block defined in IEC 61499 Part 1. (From Vyatkin, V., *IEC 61499 Function Blocks for Embedded and Distributed Control Systems Design*, 2nd edn., ISA, Durham, NC, 2012.)

4.5 Cyber-Security and Privacy Concerns

4.5.1 Nature of Cyber-Security and Privacy Concerns in Distribution Networks

One critical aspect of the information and communications technologies of smart distribution network is cyber-security, which includes physical and system security. Cyber-security covers the protection of networks and servers from unauthorized accesses and malicious attacks. It also covers the protection against compromised control and measurement units that can harm the system, physical security, secure state estimation, intrusion detection, etc.

Based on experiences gained from developed IT and telecommunication systems, it can be anticipated that the grid will be a potential target for malicious, well-equipped and well-motivated adversaries. An attack may use up the systems' own resources. These attacks can jeopardize the normal operation of the power distribution system and sometimes can be highly dangerous. Furthermore, threats such as fiddling with financial information may cause a major economical disturbance, if they are not monitored carefully.

At the consumer end, unprotected energy-related data can invade the privacy of consumers. In particular, communications signals in AMI may disclose information about where people were, when and what they were doing.

Although contemporary security technologies, such as virtual private networks (VPNs), intrusion detection systems (IDSs), public key infrastructure (PKI), anti-virus software, firewalls, etc., have well protected the IT infrastructure, before deploying they need to be proven in systems that perform the critical operations and control of smart distribution networks.

4.5.2 Basic Cyber-Security Concerns

Distribution network protection equipment and control and automation schemes use standardized data interchange and network implementations so as to make these applications interoperable among different vendors.

Accordingly, substations can now be interconnected with open networks such as corporate networks or the Internet, which use open protocols for communication. When proprietary solutions were used, they brought obscurity (as the formats and protocols are proprietary, it can be very difficult to interpret and hack into them) and isolation (as the substation network is not connected to the outside world, it can't be accessed from the outside world). However, open protocols mean that the security provided by obscurity and isolation cannot be assumed. This leaves the networks vulnerable to cyber-attacks.

With the intension of avoiding such cyber-attacks, cyber-security measures are implemented. Cyber-security considers the following in its design:

1. Confidentiality – preventing unauthorized access to information

2. Integrity – preventing unauthorized modification

3. Availability and authentication – preventing denial of service and assuring authorized access to information

4. Non-repudiation – preventing denial of an action taking place

5. Traceability and detection – monitoring and logging of activity to detect intrusion and analyse incidents

The threats to the system may be unintentional (e.g. natural disasters, human error), or intentional (e.g. cyber-attacks by hackers). Cyber-security is attainable with a range of measures such as closing down vulnerability loopholes, assuring availability, implementing adequate security processes and procedures and providing appropriate technology such as firewalls. Some common cyber-security vulnerability loopholes include

1. Indiscretions by personnel, e.g. users keeping passwords in locations visible to others

2. Bypassing of controls, such as users turning off security measures to access the system

3. Bad practice, such as users not changing default passwords, or everyone using the same password to access all substation equipment

4. Inadequate technology, where substation is not firewalled or the firewall is not tailored for substation protection

4.5.3 Solutions That Ensure Cyber-Security in Distribution Networks

Processes and procedures used in distribution network ICT infrastructure are required to assure a secure exchange of the following categories of information:

1. *Security context*: This defines information that allows users to have access to devices. It includes passwords, permissions and user credentials.

2. *Log and event management*: This includes security logs, which are stored in different IEDs.

3. *Settings*: This includes information about the IED, such as the number of used and unused ports and performance statistics.

4. *Datagrams*: Packets of information exchanged between IEDs.

The open system interconnection (OSI) 7-layer model [11,12] for peer-to-peer communication enables security measures to be applied at the Transport layer (layer 4) and the Application layer (layer 7).

At the Transport layer, the dialog between two devices is controlled. The connections between the two devices in question are established, managed and terminated at this layer by using either the transport control protocol (TCP), which ensures end-to-end reliable delivery using three-way handshaking, or the user datagram protocol (UDP), which attempts best effort delivery. Applying security at this layer is known as Transport Layer Security, and the most common protocol used here is the TLS protocol, which is also known as secure socket layer (SSL).

Security measures applied at the Transport layer guarantee confidentiality, integrity and authenticity, but because they rely on security provided by the Transport layer, this type of secure data exchange is limited to point-to-point communication. Figure 4.10 shows peer-to-peer communication between two IED nodes. The message from node A to node B is secured using the SSL protocol to prevent any possible access on its way through the network, represented by a cloud in the middle.

A more flexible solution can be implemented by applying security at the Application layer, whereby messages themselves are secured independently of the Transport layer on which they are exchanged. The application of security measures at the Application layer produces a service-oriented architecture (SOA).

FIGURE 4.10 TLS secure communication.

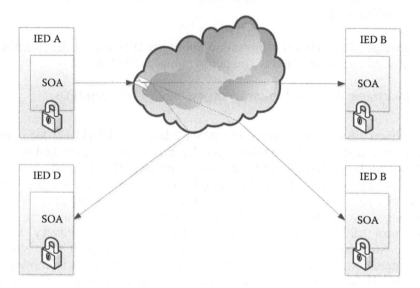

FIGURE 4.11 DPWS secure communication.

SOA is based on web services that are message oriented. Secure messages can be sent between any of the devices in the network and are not limited to point-to-point communication. SOA solutions are not prescribed to any specific profile, but the device profile for web services (DPWS) protocol provides an appropriate level of security for substation automation. Its message transfer is shown in Figure 4.11. In this case, the IED node A communicates with the IED nodes B, C and D using DPWS, which enables web profile interfacing. This is not restricted to peer-to-peer communication and enables communications among multiple nodes simultaneously and securely.

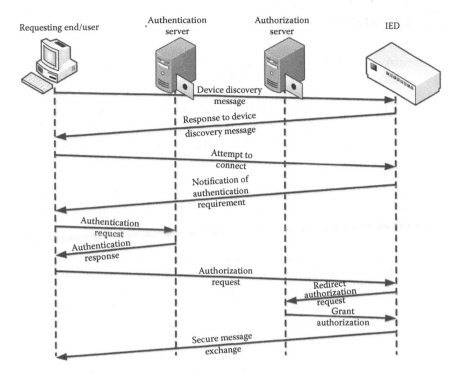

FIGURE 4.12 Implementing RBAC using an AAA server.

In addition to cyber-security on end-to-end message delivery, authentication, authorization and accounting (AAA) also has a significant role to play in granting access. Role based access control (RBAC) is a robust approach to control unintentional access by defining different classes of users granting different rights of access to different information in the devices on the network.

The following example, illustrated using Figure 4.12, shows how an AAA server is used to implement RBAC. When an IED or other networked device attached to the network is attempted access by a user through the network, an automatic device discovery process is performed. At the end of this step, the user selects the device to be connected with. The selected device requests a token from the user to prove that he/she has been authenticated by the security server. The user transfers this token to the IED. The IED then requests the user roles and credentials from the authorization server. If the roles and credentials are certified, a secure exchange of data can occur between the devices' applications [13].

☐ Questions

1. Compare wired and wireless communication technologies, and give some examples for both technologies.

2. A frequency spectrum is required to have a lowermost frequency of 2402 MHz and an uppermost frequency of 2462 MHz. Calculate how many channels having bandwidth of 5 MHz can be accommodated.

3. A process bus having an optical fibre carries three phase voltages and currents and neutral voltage and current of three bays. The bay is operating at 50 Hz.

    a. An optical fibre is used for the process bus with a statistics analysis system (SAS). The fibre is a step-index multimode fibre having a core with a refractive index of 1.5 and cladding with a refractive index of 1.485. If the length of the fibre is 500 m, what is the maximum data rate possible?

    b. According to the IEC 61850-9-2LE standard, the recommended sampling rate of each signal is 80 samples per period. Each sample needs 32 bits. Is the data rate offered by the fibre optics sufficient?

4. A communication network in which a smart meter installed in each house sends measurements to a concentrator installed at the 11/0.4 kV transformer, forming a NAN, as shown in Figure 4.4. There are 600 smart meters connected to the concentrator. Assuming 3 bytes are used to send each measurement from the smart meter, calculate the data rate required within the NAN when the sampling interval is 20 s.

5. In regard to cyber-security and privacy concerns

    a. What should be considered in its design to avoid cyber-attacks?

    b. Give some examples of typical cyber-security vulnerability loopholes.

    c. List some solutions that ensure cyber-security in distribution networks.

☐ References

1. K. Ren, Z. Li and R. C. Qiu, Guest editorial cyber, physical, and system security for smart grid, *IEEE Transactions on Smart Grid*, 2(4), 643–644, December 2011.
2. V. C. Gungor, D. Sahin, T. Kocak, S. Ergut, C. Buccella, C. Cecati and G. P. Hancke, Smart grid technologies: Communication technologies and standards, *IEEE Transactions on Industrial Informatics*, 7(4), 529–539, November 2011.
3. P. Yi, A. Iwayemi and C. Zhou, Developing ZigBee deployment guideline under WiFi interference for smart grid applications, *IEEE Transactions on Smart Grid*, 2(1), 110–120, March 2011.
4. Y. S. Kivshar and G. P. Agrawal, *Optical Solitons: From Fibers to Photonic Crystals*. Academic Press, San Diego, CA, 2003.
5. K. B. Samarakoon, Use of smart meters for frequency and voltage control, PhD thesis, School of Engineering, Cardiff University, Cardiff, UK, 2012.
6. V. C. Gungor, D. Sahin, T. Kocak, S. Ergut, C. Buccella, C. Cecati and G. P. Hancke, A survey on smart grid potential applications and communication requirements, *IEEE Transactions on Industrial Informatics*, 9(1), 28–42, February 2013.
7. Communications requirements of smart grid technologies. Department of Energy, Washington, DC, 2010. Available from: http://energy.gov/sites/prod/files/gcprod/documents/Smart_Grid_Communications_Requirements_Report_10-05-2010.pdf.
8. V. Li, F. F. Wu and J. Zhong, Communication requirements for risklimiting dispatch in smart grid, *IEEE International Conference on Communications Workshops (ICC 2010)*, Capetown, South Africa, May 2010, pp. 1–5.
9. C. Wester, M. Adamiak and J. Vico, IEC61850 protocol—Practical applications in industrial facilities, *2011 IEEE Industry Applications Society Annual Meeting (IAS)*, October 2011, pp. 1, 7, 9–13.
10. V. Vyatkin, *IEC 61499 Function Blocks for Embedded and Distributed Control Systems Design*, 2nd edn. ISA, Durham, NC, 2012.
11. R. Perlman, *Interconnections: Bridges, Routers, Switches, and Internetworking Protocols*. Addison-Wesley Professional, Boston, MA, 2000.
12. A. S. Tanenbaum, *Computer Networks*, 4th edn. Prentice Hall, Upper Saddle River, NJ, 2002.
13. Alstom Grid, *Network Protection & Automation Guide*. Alstom Grid Worldwide Contact Centre, UK, May 2011.

CHAPTER 5

Power Electronics in Distribution Systems

5.1 Introduction

With the advance of the smart grid, there will be an increasing shift from

- Centralized to distributed generation

- Conventional energy sources to renewables

- Manual operation to automation

- Call centres to automatic actions

- Passive to active distribution networks

- Customers to prosumers

- AC networks to DC networks

In the process of successful implementation of all these paradigm shifts, power electronics will play an important role. In a manner similar to motor neurons in the brain taking decisions and passing them to muscles for certain tasks, smart grid computers take decisions and pass them through communication infrastructure to power electronic equipment for action.

5.2 Grid Interface for DERs

Distributed energy resources (DERs) can include micro-turbines, fuel cells, wind power, solar power and several forms of energy storage. As electrical vehicles (EVs) can act as a load or storage, they are also categorized as DERs.

5.2.1 Wind Turbines

The operation of different wind turbine technologies is discussed in Chapter 2. As discussed, most of the present-day wind turbines are of variable speed and have a range of designs as summarized in Table 5.1.

As discussed in Section 2.4.3.1, a DFIG wind turbine generator uses a wound-rotor induction generator with slip rings to take power into or out of the rotor winding. The rotor winding is fed through a variable-frequency power converter, typically based on two voltage source converters (VSCs) linked by a DC bus. When it is required to increase the speed of the turbine above synchronous speed to achieve maximum power, real power is injected to the network by the rotor-side converter. On the other hand, in order to decrease the speed below the synchronous speed, real power is absorbed by the rotor from the rotor-side converter.

Example 5.1

A 1.65 MW, 4-pole wound rotor induction generator is employed for a DFIG wind turbine connected to a 50 Hz system. The rotor diameter of the turbine is 77 m. The gear box ratio is 1:115. The turbine operates at the optimum tip speed ratio of $\lambda_{opt} = 7$.

a. At what wind speed does the rotational speed of the generator become synchronous speed?
b. If the wind speed is 10 m/s, what should be the frequency of the voltage that should be applied to the rotor to maintain tip speed ratio at λ_{opt}? Should this voltage be in phase or in anti-phase with the rotor-induced voltage?
c. If the wind speed is 5 m/s, what should be the frequency of the voltage that should be applied to the rotor to maintain tip speed ratio at λ_{opt}? Should this voltage be in phase or in anti-phase with the rotor-induced voltage?
d. What should be the rating of the converter required to achieve this operation?

TABLE 5.1 Variable-Speed Wind Turbine Technologies

Technology	Generator	Gear Box	Stator Connection	Rotor Connection
Doubly fed induction generator (DFIG)	Wound-rotor induction generator	Three-stage gear box	Through a transformer to the wind farm collector network	Through back-to-back power electronic converters to the stator transformer
Full-scale frequency converter (FSFC) based	Synchronous (cylindrical rotor) or squirrel cage induction generator	Single-stage gear box	Through back-to-back power electronic converters to the wind farm collector network	A synchronous generator rotor is fed with DC; whereas there is no connection to the rotor of a cage induction generator
	Permanent magnet synchronous	None	Through back-to-back power electronic converters to grid transformer	None

Answer

a. Since for a synchronous generator $\omega = 120 \; f/p$, the synchronous speed of the generator $= 120 \times 50/4 = 1500$ rev/min $= 157.08$ rad/s.

Since the gearbox ratio is 1:115, the corresponding low speed shaft speed $= 157.08/115 = 1.3659$ rad/s.

Then from $\lambda = \dfrac{\omega R}{U}$,

$$7 = \frac{1.3659 \times 77/2}{U}$$

$\therefore \; U = 7.51$ m/s

b. When $U = 10$ m/s,

$$7 = \frac{\omega \times 77/2}{10}$$

$\therefore \omega = 1.818$ rad/s

The corresponding high speed shaft speed $= 1.818 \times 115 = 209.09$ rad/s $= 1996.67$ rev/min

Slip frequency $= (1996.67 - 1500)/1500 = 0.33$

As the generator runs at super-synchronous speed, power is absorbed from the rotor. Therefore the injected voltage should be in anti-phase with the rotor-induced voltage.

c. When $U = 5$ m/s,

$$7 = \frac{\omega \times 77/2}{5}$$

$\therefore \omega = 0.909$ rad/s

The corresponding high speed shaft speed $= 0.909 \times 115 = 104.55$ rad/s $= 998.34$ rev/min

Slip frequency $= (998.34 - 1500)/1500 = -0.33$

As the generator runs at sub-synchronous speed, power is injected into the rotor. Therefore, the injected voltage should be in-phase with the rotor-induced voltage.

d. The rating of the converter can be derived using the power balance. Mechanical power (P_m) extracted from the rotor plus or minus power through the slip rings (P_r) minus rotor Cu loss is transferred to the stator through the air gap. A part of this power is lost as stator Cu losses and core losses. The remaining power (P_e) is transferred to the grid.

Neglecting losses, the air gap power $= P_m \pm P_r = P_e$

If T is the torque,

$$T\omega_s = T\omega_r \pm P_r$$

$$P_r = T(\omega_s - \omega_r) = Ts\omega_s = sP_e$$

Rating of the converter $= 0.33 \times 1650 = 544.5$ kVA

Full-scale frequency converter (FSFC)-based wind turbines have four main forms (Figure 5.1). Some employ a cylindrical pole synchronous generator or an induction generator whereas some employ a multi-pole, permanent magnet synchronous generator. The multi-pole design enables the connection of the generator shaft directly to the turbine shaft thus allowing low-speed operation of the generator. As all of the power from the turbine goes

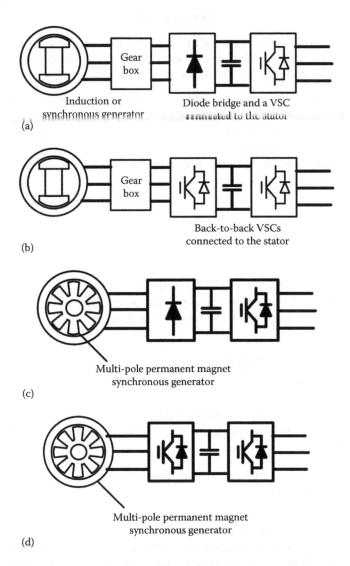

FIGURE 5.1 Full Scale Frequency Converter wind turbines: (a) geared with a diode bridge and VSC, (b) geared with back-to-back VSCs, (c) gearless with a diode bridge and VSC, and (d) gearless with back-to-back VSCs.

through the VSCs, the dynamic operation of the electrical generator is effectively isolated from the power grid. The electrical frequency of the generator may vary as the wind speed changes, while the grid frequency remains unchanged, thus enabling variable-speed operation of the wind turbine.

Example 5.2

A 1.5 MW, 88-pole permanent magnet synchronous generator is employed for a gearless FSFC wind turbine. The blade diameter of the turbine is 82 m. The optimum tip speed ratio is $\lambda_{opt} = 9.2$.

a. Calculate the generator-side converter frequency required to maintain the tip speed ratio at λ_{opt} if the wind speed is (i) 12 m/s (rated wind speed), (ii) 6.5 m/s and (iii) 3.5 m/s (cut-in wind speed).
b. What should be the rating of the converter required to achieve this above operation?

Answer

a. At a wind speed of 12 m/s from $\lambda = \dfrac{\omega R}{U}$,

$$9.2 = \frac{\omega \times 82/2}{12}$$

$\therefore \omega = 2.69$ rad/s $= 25.71$ rev/min

Since for a synchronous generator $\omega = 120\ f/p$,

The frequency of the converter voltages $= 25.71 \times 88/120 = 18.85$ Hz

At a wind speed of 6.5 m/s from $\lambda = \dfrac{\omega R}{U}$,

$$9.2 = \frac{\omega \times 82/2}{6.5}$$

$\therefore \omega = 1.46$ rad/s $= 13.93$ rev/min

Frequency of the converter voltages $= 13.93 \times 88/120 = 10.22$ Hz

At a wind speed of 3.5 m/s from $\lambda = \dfrac{\omega R}{U}$,

$$9.2 = \frac{\omega \times 82/2}{3.5}$$

$\therefore \omega = 0.785$ rad/s $= 7.5$ rev/min

Frequency of the converter voltages $= 7.5 \times 88/120 = 5.5$ Hz

b. The rating of the converter should be equal to the rating of the generator, that is 1.5 MVA.

Figure 5.2 shows a VSC used for variable-speed wind turbines. Depending on the design, one or two VSCs are employed. The switches S_1 to S_6 shown could be insulated gate bipolar transistors (IGBTs) or metal oxide field effect transistors (MOSFETs). Both switches in one limb are switched in opposition with a very small time delay. Therefore, one switching function is enough to control both switches in a limb. The time delay is required to prevent the failure of the converter during switch turn on and turn off transitions.

A number of different switching techniques can be used to turn on and off switch pairs (S_1 and S_2), (S_3 and S_4) and (S_5 and S_6). One of the commonly used switching techniques, sine-triangular pulse width modulation (PWM), is shown in Figure 5.3. Here, a modulating signal with the same frequency and phase angle as the desired output wave form of the VSC is compared with a high-frequency triangular carrier signal to generate switching signals. Figure 5.3 shows the switching signals of switch S_1, that of switch S_2 is complementary to the switching signal of S_1. Switches S_3 and S_5 have the same switching signal as that of S_1 but are phase-shifted by 120° and 240°.

Another switching technique employed in VSCs is the space vector PWM (SVPWM), in which switching vectors are generated by the states of the converter switches. Three switching legs, (S_1 and S_2), (S_3 and S_4) and (S_5 and S_6), each having two states (ON or OFF), allow the converter to produce eight possible switching states, as given in Table 5.2. 'H' and

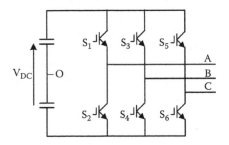

FIGURE 5.2 Six-pulse VSC.

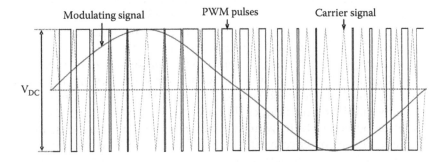

FIGURE 5.3 PWM switching strategy.

TABLE 5.2 SVPWM Switching Vectors

S_1 and S_2	S_3 and S_4	S_5 and S_6	V_{AO}	V_{BO}	V_{CO}	Switching Vector
L	L	L	$-V_{DC}/2$	$-V_{DC}/2$	$-V_{DC}/2$	V_0
H	L	L	$+V_{DC}/2$	$-V_{DC}/2$	$-V_{DC}/2$	V_1
H	H	L	$+V_{DC}/2$	$+V_{DC}/2$	$-V_{DC}/2$	V_2
L	H	L	$-V_{DC}/2$	$+V_{DC}/2$	$-V_{DC}/2$	V_3
L	H	H	$-V_{DC}/2$	$+V_{DC}/2$	$+V_{DC}/2$	V_4
L	L	H	$-V_{DC}/2$	$-V_{DC}/2$	$+V_{DC}/2$	V_5
H	L	H	$+V_{DC}/2$	$-V_{DC}/2$	$+V_{DC}/2$	V_6
H	H	H	$+V_{DC}/2$	$+V_{DC}/2$	$+V_{DC}/2$	V_7

'L' represent the switching leg's status. When phase-a upper switch is ON then S_a = 'H' and when it is OFF then S_a = 'L'.

The switching vectors given in Table 5.2 are the voltage vectors produced by the converter outputs for corresponding switching states. The switching vectors V_0 and V_7 are null vectors. The remaining six switching vectors V_1 to V_6 are equally displaced by 60°. In SVPWM generation, the sequence of the switching vectors is selected in such a way that only one leg is switched to move from one switching vector to the next. More details about this switching technique can be found in [1,2]. The main advantage of this switching technique is that it can easily be implemented in a digital environment.

Other than the two back-to-back VSCs employed to extract maximum power from wind, more power electronic circuits are employed in wind turbines to provide other services demanded by present-day grid codes. One of the critical circuits is that required to fulfil the fault ride through (FRT) capability. During a network fault, as the power transferred to the grid drops, the incoming power from the turbine should be absorbed by an external means; otherwise the generator speed will increase and it may trip on its over-speed protection. Figure 5.4 shows some of the commonly employed techniques that are used to dissipate the incoming mechanical power. In DFIG wind turbines, as shown in Figure 5.4a and b, during a fault the power converters are blocked and a crowbar is switched on. The crowbar resistor absorbs excess power thus providing the FRT. Figure 5.4c shows a chopper which is employed to absorb excess energy.

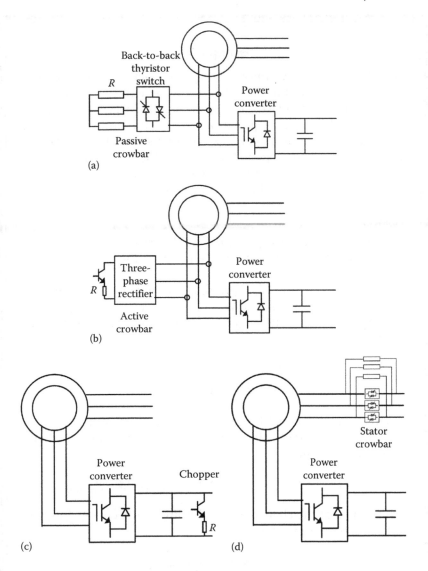

FIGURE 5.4 FRT techniques employed in DFIG wind turbines: (a) rotor passive crowbar, (b) rotor active crowbar, (c) rotor chopper and (d) stator crowbar.

The chopper is triggered based on the DC link voltage. When the DC link voltage is above the threshold value, the chopper is turned on, and after a while it is turned off. This way the DC link voltage can be maintained at the maximum limit. As shown in Figure 5.4d, a stator crowbar can be used to divert the power to a resistor bank. However under normal operation the conduction through the back-to-back thyristor pair incurs extra losses.

Example 5.3

A 1.65 MW, 4-pole wound-rotor induction generator is employed for a DFIG wind turbine. When the generator is producing power at its rated capacity, a fault occurs on the grid side and the power transferred to the grid drops to 0.2 pu (with respect to the machine rating). Calculate the power that should be absorbed by the crowbar resistor to maintain the machine speed within 1% of its rated speed. Assume that the fault is cleared within 120 ms. The inertia constant of the generator is 2.5 s.

Answer

The change in speed is governed by $d\omega/dt = (1/J)\Delta T$, where J is the inertia and ΔT is the difference between the mechanical and electrical torques.

The inertia can be expressed per unit as an H constant where $H = \left((1/2)J\omega_s^2\right)/S_{rated}$ (Ws/VA). In this equation, S_{rated} is the base MVA and ω_s is the angular velocity (rad/s) at synchronous speed. Thus

$$\frac{d\omega}{dt} = \frac{\omega_s^2}{2HS_{rated}}\Delta T = \frac{\omega_s\Delta P}{2H}$$

where $(\omega_s\Delta T)/S_{rated} = \Delta P$ is the per unit power difference.

During the fault, since the rotor conductor is blocked and crowbar is activated,

$$\Delta P = P_m - P_e - P_{crowbar} = 1 - 0.2 - P_{crow\_bar} = 0.8 - P_{crowbar}$$

Assuming that the change in speed is linear, the change in speed within 120 ms is given by

$$\Delta\omega = \frac{\omega_s\left[0.8 - P_{crowbar}\right]}{2H}\Delta t$$

In order to limit the final speed to 1%, i.e. $(\Delta\omega/\omega_s) = 0.01$

$$0.01 = \frac{\left[0.8 - P_{crowbar}\right]}{2 \times 2.5} \times 0.12$$

$P_{crowbar} = 0.38$ pu $= 1650 \times 0.38 = 632.5$ kW

In FSFC wind turbine, the commonly employed FRT technique is a chopper connected to the DC link of the back-to-back VSCs. If the power losses in the converters are neglected, the power delivered to the grid is equal to the power generated before the fault occurs. During the fault, the grid power drops and the capacitor absorbs excess power. As the capacitor voltage increases, the chopper is turned on, dissipating the excess power in the resistor.

When the chopper is OFF $V_{DC} \times C(dV_{DC}/dt) = P_{Generated} - P_{Supplied}$. Integrating this equation, the following equation is obtained:

$$\frac{1}{2}CV_{DC}^2 = \left[P_{Generated} - P_{Supplied} \right] t + k \qquad (5.1)$$

where k is the constant of integration

At $t = 0$ s, assume that $V_{DC} = V_0$: then $k = (1/2)CV_0^2$. Then

$$\frac{1}{2}C\left[V_{DC}^2 - V_0^2 \right] = \left[P_{Generated} - P_{Supplied} \right] t \qquad (5.2)$$

When the chopper is ON, the equation that governs the DC link voltage is

$$V_{DC} \times C \frac{dV_{DC}}{dt} = P_{Generated} - P_{Supplied} - P_{Chopper} \qquad (5.3)$$

Example 5.4

A 1.5 MW synchronous generator is employed for a gearless FSFC wind turbine. When the generator is producing power at its rated capacity, a fault occurs on the grid side and the power transferred to grid drops to 0.2 pu (with respect to the machine rating). From the time of the fault, at what time should the chopper be turned on to maintain the DC link voltage within 1% of its pre-fault value? The DC link capacitor is 500 μF.

Answer

When the DC link voltage is 1% above its pre-fault value, $V_{DC}/V_0 = 1.01$.

From Equation 5.2,

$$\frac{1}{2}C\left[V_{DC}^2 - V_0^2 \right] = \left[P_{Generated} - P_{Supplied} \right] t$$

$$0.5 \times 500 \times 10^{-6} \times \left[1.01^2 - 1 \right] = \left[1 - 0.2 \right] \times t$$

$$t = 6.28 \ \mu s$$

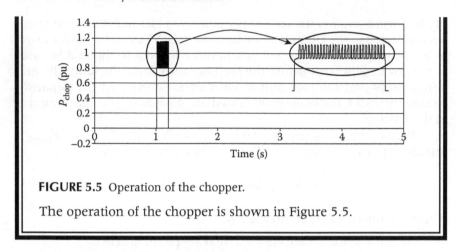

FIGURE 5.5 Operation of the chopper.

The operation of the chopper is shown in Figure 5.5.

5.2.2 Photovoltaic Generators

The typical components of a photovoltaic (PV) generator are shown in Figure 5.6. The components shown in solid lines appear in almost all grid-connected and off-grid systems. In grid-connected applications, the inverter maintains the DC link voltage corresponding to the AC side voltage. Therefore the DC–DC converter 1 is a boost converter which matches the PV array voltage to the DC link voltage. As power always flows from the PV array to the grid, this can be a unidirectional converter. If a battery is present, then during the time the battery is charging, the power flows from the DC link to the battery. On the other hand, when the battery is discharging, the power flows from the battery to the load/grid. Therefore, a bidirectional DC–DC converter is required for DC–DC converter 2.

Table 5.3 shows different DC–DC converters that can be used for PV systems. The relationships between the output and input voltages are also given [3,4].

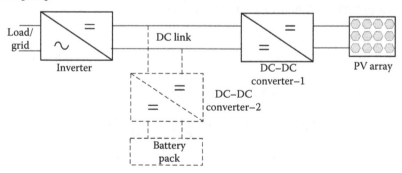

FIGURE 5.6 Typical components of a PV generator.

TABLE 5.3 Different Types of DC/DC Converter Circuits (in All Figures)

Boost Converter

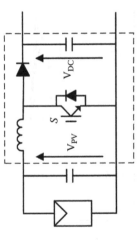

$V_{DC} = V_{PV}/(1 - D)$
where D is the duty ratio of the switch S; suitable for converter 1 but not for converter 2

Buck-boost converter

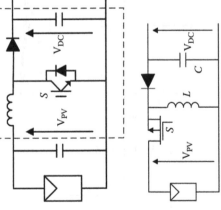

$V_{DC} = [D/(1 - D)]V_{PV}$
where D is the duty ratio of the switch S; suitable for both converter 1 and converter 2

Full-bridge DC–DC converter

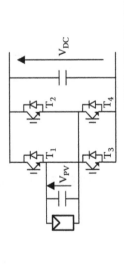

For bipolar and unipolar switching; $V_{DC} = (2D - 1)V_{PV}$
where D is the duty ratio of the switch pair T_1 and T_4 and the duty ratio of T_2 and T_3 is $(1 - D)$; suitable for both converter 1 and converter 2

Example 5.5

A PV module with 36 cells and with the characteristic shown in Figure 5.7 is connected to a DC–DC converter as shown in Figure 5.8. The DC–DC converter uses PWM with bipolar

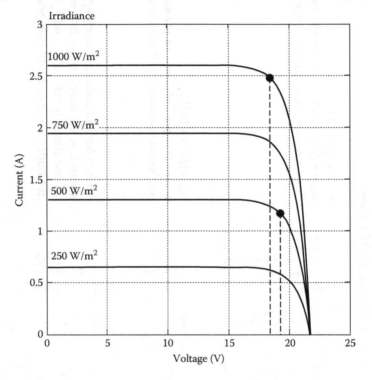

FIGURE 5.7 I–V characteristics of the PV module.

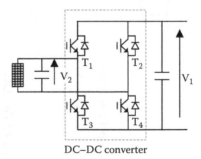

DC–DC converter

FIGURE 5.8 Full-bridge DC–DC converter.

switching [8]. If $V_1 = 25$ V, what should be the duty ratio D to extract maximum power from the PV panel when the irradiance is (a) 1000 W/m² and (b) 500 W/m².

Answer

a. At 1000 W/m², in order to extract maximum power, the voltage across the module should be 18 V.
   Then $18 = (2D-1) \times 25$ ∴ $D = 0.86$
b. At 500 W/m², in order to extract maximum power, the voltage across the module should be 19 V.
   Then $19 = (2D-1) \times 25$ ∴ $D = 0.88$

The commonly used inverter for PV application is the H-bridge inverter shown in Figure 5.8. Two switching techniques, namely bipolar switching and unipolar switching, are used to turn ON and OFF the switches in the inverter. In bipolar switching, the switch pair (T_1 and T_4) and (T_2 and T_3) are turned on in a complementary manner by comparing a triangular waveform with a modulating signal, as shown in Figure 5.3. In the case of unipolar switching, two modulating signals are used; one is same as the desired waveform (V_{mod}) and the other is a 180° phase-shifted waveform ($-V_{mod}$). When the amplitude of the triangular waveform is less than that of the modulating signal V_{mod}, then T_1 is ON; otherwise, T_3 is ON. Similarly, when the amplitude of the triangular wave form is less than that of the modulating signal $-V_{mod}$, then T_2 is ON; otherwise, T_4 is ON. Figure 5.9 shows the inverter output with the unipolar and bipolar PWM techniques.

5.2.3 Electric Vehicles (EV)

EVs employ a number of power electronic conversion stages, as shown in Figure 5.10. Most traction applications require motor(s) with a capacity 75–300 kW. In order to minimize losses associated with such a motor, it should be driven at a fairly higher voltage (compared to current voltages of 12–24 V). Typically, a DC link voltage of 250–400 V is utilized for electric cars. For example, the Nissan Leaf employs an 80 kW motor with a battery of a nominal voltage of 360 V. Motors such as brushless DC motors, induction motors and permanent magnet synchronous motors that can operate at high voltage and provide good overall efficiency are selected for vehicle traction applications.

The traction drive connected to the drive motor needs to be operating in both quadrants. When the vehicle moves, power flows from the

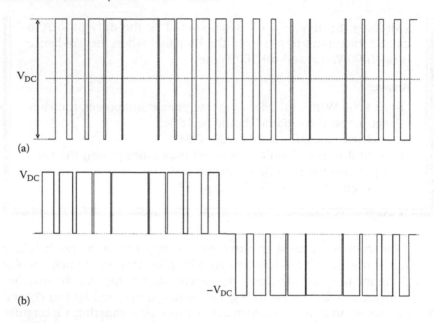

FIGURE 5.9 Full-bridge inverter output: (a) output waveform with bipolar switching and (b) output waveform with unipolar switching.

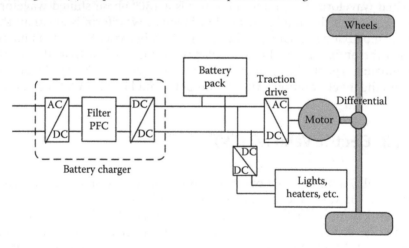

FIGURE 5.10 Components of an electric car.

DC link to the motor. During regenerative braking, power flows from the motor to the DC link. This adjustable traction drive system has a controller which basically controls the motor speed by an input speed reference signal that can be adjusted by the position of the accelerator paddle.

Example 5.6

An EV is driven by a brushless DC motor. The parameters of the motor are as follows:

Efficiency (η) = 94%
Field constant ($k\phi$) = 0.1 Vs/rad
Armature resistance = 0.1 Ω

The weight of the EV is 1500 kg and the motor is connected through a gearbox with a ratio (G) 6:1. The following data are given as the design operating conditions (based on [5]):

EV rated speed (v) = 80 km/h
Aerodynamic drag coefficient (C_{drag}) = 0.3
Tire rolling resistance coefficient (C_{rr}) = 0.013
Density of dry air (ρ) = 1.2 kg/m^2
Angle of the driving surface (α) = 10°
Front area (A) = 2.1 m^2
Wheel radius = 0.2794 m
Zero head wind

a. What is the power rating of the required electrical motor?
b. If a 4-quadrant converter is used to drive the motor, what should be the duty ratio of the switches at the design conditions given above? The DC link voltage is 350 V.

Answer

a. Figure 5.11 shows the EV on the driving surface. The forces shown are as follows:

F_I is the Inertial force of the EV.
F_g is the Gravitational force of the EV.
F_{rr} is the Rolling resistance force of the wheels.
F_w is the Force due to wind resistance.
F_t is the Traction force of the EV.

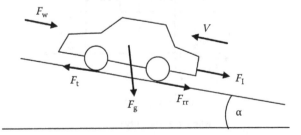

FIGURE 5.11 Forces acting on an EV.(From Rahman, K.M. and Ehsani, M., Performance analysis of electric motor drives for electric and hybrid electric vehicle applications, *IEEE Power Electronics for Transportation*, 1996, pp. 49–56.)

In steady state [6],

$$F_t = F_g \sin\alpha + F_{rr} + F_I + F_w$$

$$= Mg\sin\alpha + C_{rr}Mg\cos\alpha + \frac{1}{2}\rho C_{drag}A\left(v + v_{wind}\right)^2$$

$$= 1500\times9.81\times\sin10^0 + 0.013\times1500\times9.81\times\cos10^0$$

$$+ 0.5\times1.2\times0.3\times2.1\times\left(80\times\frac{1000}{3600}\right)$$

$$= 2752.02\,N$$

Since the radius of wheel is $r = 0.2794$ m, the corresponding torque $= \tau_t = F_t r = 2752.02\times0.2794 = 768.91$ Nm.

The shaft torque of the electric motor

$$\tau_s = \eta\frac{\tau_t}{G} = 0.94\times\frac{768.91}{6} = 120.46\,Nm.$$

The shaft angular velocity of the electric motor (ω)

$$= G\frac{v}{r} = 6\times\frac{\left(80\times1000/3600\right)}{0.2794} = 477.21\,rad/s$$

The shaft power of the electric motor $= P_s = \tau_s\omega = 120.46\times477.21 = 57.48$ kW.

b. The motor is connected as shown in Figure 5.12.

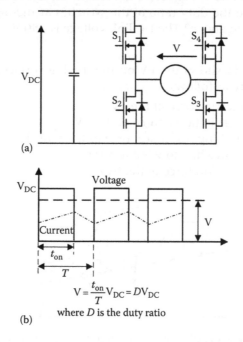

(a)

(b)

$$V = \frac{t_{on}}{T}V_{DC} = DV_{DC}$$

where D is the duty ratio

FIGURE 5.12 Motor drive system (a) Motor connected to the full bridge converter. (b) Current and voltage through the motor.

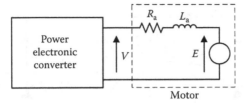

FIGURE 5.13 Equivalent circuit of the motor drive system shown in Figure 5.12.

The equivalent circuit of the motor drive circuit shown in Figure 5.12 is shown in Figure 5.13.

From the equivalent circuit,

$$V = R_a i_a + L_a \frac{di_a}{dt} + E$$

In steady state,

$$V = R_a i_a + E$$

For the DC motor,

$$E = k\phi\omega$$

$$\tau = k\phi i_a$$

Combining these equations, the following equation was obtained for the terminal voltage of the motor:

$$V = R_a \frac{\tau}{k\phi} + k\phi\omega$$

$$= 0.1 \times \frac{120.46}{0.1} + 0.1 \times 477.21 = 168.18\,\mathrm{V}$$

Therefore, the duty ratio of the switches should be = 168.18/350 = 0.48.

If the motor used for the traction drive is an induction motor or a permanent magnet synchronous motor, then a three-phase inverter, as shown in Figure 5.2, is used.

The battery charger shown in Figure 5.10 could be an on-board charger or could be located in the charging station. If grid-to-vehicle (G2V) is the only option available in the charger, then the unidirectional charger shown in Figure 5.14 is employed. This charger consists of an AC–DC converter, a power factor correction (PFC) circuit and a DC–DC converter.

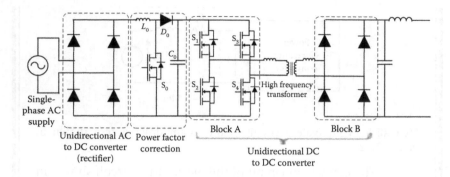

FIGURE 5.14 Unidirectional battery charger from G2V.

The PFC circuit is a boost converter. During the time the MOSFET S_0 is ON, energy is stored in the inductor, L_0, and charging current is supplied by capacitor C_0 (stored energy). When S_0 is OFF, the inductor and the source supply the charging current and also transfer energy to the capacitor (C_0). The MOSFET is switched ON using a hysteresis current controller where the reference is in phase with the supply voltage.

Block A of the DC–DC converter converts DC into high-frequency AC and Block B converts high-frequency AC into DC. As a high-frequency transformer is used to couple Block A and B, its size becomes small compared to a 50 Hz transformer. The DC–DC converter creates the necessary charge profile for the battery.

If both G2V and vehicle to grid (V2G) options are required, the AC–DC converter should be bidirectional. Depending on the size of the charger, one of the AC–DC bidirectional converters shown in Figure 5.15 is employed for the battery charger.

5.3 Devices for Power Quality Improvement

5.3.1 Dynamic Voltage Restorer

A dynamic voltage restorer (DVR) is a series-connected device that injects a set of three-phase AC output voltages in series with the distribution feeder voltages. It is often used to restore voltage sags. A DVR employs a PWM inverter and is connected to the line through a series transformer, as shown in Figure 5.16. The amplitude and phase of the injected voltage to the feeder can be changed by the inverter.

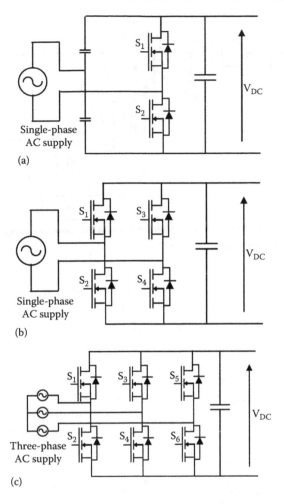

FIGURE 5.15 Bidirectional AC–DC converters: (a) single-phase half-bridge AC–DC converter, (b) single-phase full-bridge AC–DC converter and (c) three phase AC–DC converter.

Three operating modes are available:

1. *Standby mode*: The PWM inverter is blocked and the secondary side of the series transformer is short-circuited through a crowbar or by one leg of the inverter.

2. *Compensation mode*: If there is a voltage sag on the distribution feeder, then voltage is boosted by injecting an appropriate voltage.

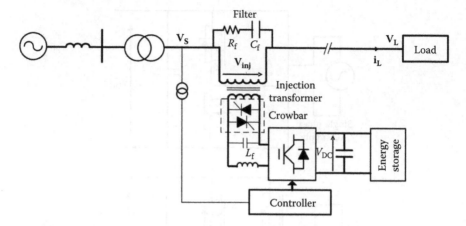

FIGURE 5.16 Power circuit of a DVR.

3. *Fault protection mode*: If there is a fault in the distribution feeder to which the DVR is connected and the fault current is greater than a certain maximum, then the crowbar is turned on to protect the inverter.

In the compensation mode, two compensation techniques are normally employed: reactive power compensation [7–9] and active power compensation [9,10]. Figure 5.17a shows the phasor diagram for the reactive power compensation. As shown, the injected voltage V_{inj} is in quadrature

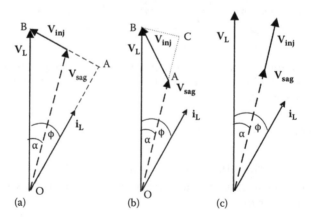

FIGURE 5.17 Phasor diagrams for different compensation modes: (a) reactive power compensation, (b) active power compensation: pre-sag and (c) active power compensation: in-phase.

with the load current $\mathbf{i}_L$. From the phasor diagram, the following equation can be obtained:

$$
\begin{aligned}
\left|\mathbf{V}_{sag}\right| &= \sqrt{\left(\left|\mathbf{V}_L\right|\cos\phi\right)^2 + \left(\left|\mathbf{V}_L\right|\sin\phi - \left|\mathbf{V}_{inj}\right|\right)^2} \\
&= \sqrt{\left|\mathbf{V}_L\right|^2 + \left|\mathbf{V}_{inj}\right|^2 - 2\left|\mathbf{V}_{inj}\right|\left|\mathbf{V}_L\right|\sin\phi}
\end{aligned}
\tag{5.4}
$$

where

$\mathbf{V}_L$ is the load voltage before the sag
$\mathbf{V}_{sag}$ is the load voltage during the sag
$\mathbf{V}_{inj}$ is the injected voltage of the DVR
ϕ is the load angle

Reactive power compensation is only effective to voltage sags that follow the trajectory AB. Furthermore, this compensation method does not provide dynamic compensation for sags. Therefore, active power compensation techniques are usually employed. In the pre-sag compensation shown in Figure 5.17b, the injected voltage restores the magnitude and phase angle of the voltage at the load to pre-sag voltage, i.e. to $\mathbf{V}_L$. In order to track the load voltage, a phase-locked loop (PLL) is employed. The PLL tracks the positive sequence component of the supply voltage in normal operation but freezes the phase angle during a sag. In in-phase compensation the injected voltage restores the magnitude of the load voltage to pre-sag value, i.e. to $|\mathbf{V}_L|$. This technique demands a fast PLL.

Example 5.7

A part of a distribution network is shown in Figure 5.18. Under normal operating conditions, the voltage of busbar C is $1\angle 0°$ pu. Due to a large load connected to busbar C, its voltage drops and the voltage phasor changes by 5°. If the DVR is capable of injecting a voltage of 0.1 pu, what is the minimum possible voltage sag that can be compensated by a DVR connected at busbar B, if the compensation technique used is

a. Reactive power compensation
b. Active power compensation: pre-sag
c. Active power compensation: in-phase

Answer

a. Reactive power compensation
From Equation 5.4,

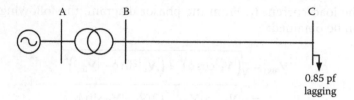

FIGURE 5.18 Distribution network for Example 5.7.

$$\left|\mathbf{V}_{sag}\right|_{min} = \sqrt{\left|\mathbf{V}_L\right|^2 + \left|\mathbf{V}_{inj}\right|^2 - 2\left|\mathbf{V}_{inj}\right|\left|\mathbf{V}_L\right|\sin\phi}$$
$$= \sqrt{1 + 0.1^2 - 2\times1\times0.1\times\sin\left(\cos^{-1}0.85\right)} = 0.95$$

b. Active power compensation: pre-sag
   From Figure 5.17b

$$\left|\mathbf{V}_{sag}\right| = OC - AC = OC - \sqrt{AB^2 - BC^2}$$
$$= \left|\mathbf{V}_L\right|\cos\alpha - \sqrt{\left(\left|\mathbf{V}_{inj}\right|\right)^2 - \left(\left|\mathbf{V}_L\right|\sin\alpha\right)^2}$$
$$= 1\times\cos5° - \sqrt{\left(0.1\right)^2 - \left(1\times\sin5°\right)^2}$$
$$= 0.947\,pu$$

c. Active power compensation: in-phase
   From Figure 5.17c: $\left|\mathbf{V}_{sag}\right| = 1 - 0.1 = 0.9\,pu$

5.3.2 Static Compensator

A static compensator (STATCOM) is a shunt device that also employs a PWM converter as shown in Figure 5.19. By injecting reactive power, it can compensate for voltage dips in distribution circuits. It can also improve the current quality in the distribution line by absorbing harmonic currents generated by the load.

The PWM converter is controlled to produce a fundamental voltage in phase with the power system voltage. As shown in the phasor diagram in Figure 5.19a, when the fundamental converter voltage is greater than the AC system voltage, a leading current is drawn from the STATCOM, thus it produces reactive power (Figure 5.19c). On the other

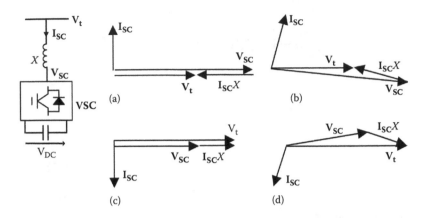

FIGURE 5.19 Operation of a STATCOM: (a) $V_{SC} > V_t$: STATCOM generates Q, (b) V_{SC} lags V_t: P flows from AC to DC side, (c) $V_{SC} < V_t$: STATCOM absorbs Q and (d) V_{SC} leads V_t: P flows from DC to AC side.

hand, when the converter voltage is lower than the AC system voltage, STATCOM absorbs reactive power. Under steady-state operating conditions, no real power exchange takes place. Therefore, a DC capacitor which provides the commutating voltage is sufficient.

As shown in Figure 5.19b, if the converter terminal voltage lags the AC supply voltage, then power flows from the AC side to the DC side thus charging the capacitor. A small lagging angle is used to control the reactive power generated by the STATCOM. If the converter terminal voltage leads the AC supply voltage, then power flows from the DC side to the AC side (Figure 5.19d) thus discharging the capacitor. A small leading angle is used to control the reactive power absorbed by the STATCOM.

The application of a shunt device, such as a STATCOM, for the mitigation of sags has advantages when compared to that of a series device, as the shunt devices can simultaneously be used for steady-state voltage control and power oscillation damping. However, as the shunt device can compensate only for a small voltage dip when compared to a series device, the effectiveness for sag mitigation is limited [11].

Figure 5.20a shows how a STATCOM can be employed for sag mitigation, and the corresponding phasor diagram is shown in Figure 5.20b. In Figure 5.20b, the full lines indicate the phasor relationship before the sag. Assuming that the sag is caused by an increase in load current with the same power factor, without the STATCOM injecting any current, the load voltage would reduce to V_{sag}. If the STATCOM injects a current I_{SC}, then the voltage can be restored to V_{Com}, with a magnitude more or less same as the pre-sag value.

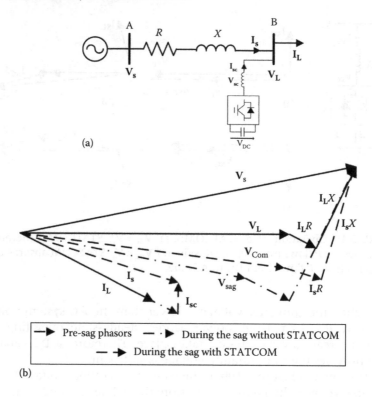

(a)

(b)

FIGURE 5.20 A STATCOM for sag mitigation: (a) STATCOM connection and (b) Phasor diagram.

Example 5.8

A part of a distribution network is shown in Figure 5.21. The voltage at busbar A is $1\angle 0°$. When the motor connected to busbar C starts, it draws a current three times the rated current at 0.6 pf lagging for 50 ms. As other sensitive loads are connected

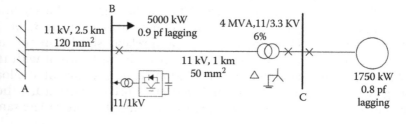

FIGURE 5.21 A network for Example 5.8.

to busbar B, its voltage should be maintained above 0.9 pu. Calculate the minimum capacity of the STATCOM to achieve this. Data for the lines are as follows:

$$50 \text{ mm}^2: R = 0.494 \text{ } \Omega/\text{km}, X = 0.412 \text{ } \Omega/\text{km}$$
$$120 \text{ mm}^2: R = 0.196 \text{ } \Omega/\text{km}, X = 0.361 \text{ } \Omega/\text{km}$$

Answer

Rated current of the motor $= \dfrac{1750 \times 10^3}{\sqrt{3} \times 3.3 \times 10^3 \times 0.8} = 382.7 \text{ A}$

Starting current $= 3 \times 382.7 = 1148.14 \text{ A}$

As the power factor is 0.6, the starting current of the motor

$$= 1148.14 \angle -\cos^{-1}0.6 = 1148.14 \angle -53.13°$$

Current drawn by the load connected to busbar

$$B = \dfrac{5000 \times 10^3}{\sqrt{3} \times 11 \times 10^3 \times 0.9} = 291.6 \text{ A}$$

As the power factor is 0.9, the current drawn by the load

$$= 291.6 \angle -\cos^{-1}0.9 = 291.6 \angle -25.84°$$

Selecting $S_{\text{base}} = 1000$ kVA, then $I_{\text{base}} = \dfrac{1000 \times 10^3}{\sqrt{3} \times 11 \times 10^3} = 52.49 \text{ A}$

If STATCOM injects a current of $I_Q \angle 90°$ pu, then the total current through line section AB

$$= \dfrac{1148.14 \angle -53.13°}{52.49} \times \dfrac{3.3}{11} + \dfrac{291.6 \angle -25.84°}{52.49} + I_Q \angle 90°$$
$$= 6.56 \angle -53.13° + 5.56 \angle -25.84° + I_Q \angle 90° \text{ pu}$$
$$= 8.94 + j(I_Q - 7.67)$$

Z_{base} on 11 kV $= \dfrac{11^2}{1} = 121 \text{ } \Omega$

Since Line AB is 120 mm², its impedance

$$= \dfrac{(0.196 + j0.361) \times 2.5}{121} = 0.0041 + j0.0075 \text{ pu}$$

The voltage drop across line section AB should be less than 0.1 pu, i.e.

$$\left| \left[8.94 + j\left(I_Q - 7.67 \right) \right] \times \left(0.0041 + j0.0075 \right) \right| \le 0.1$$

$$\left| \begin{array}{l} \left[8.94 \times 0.0041 - 0.0075 \times \left(I_Q - 7.67 \right) \right] \\ + j\left[8.94 \times 0.0075 + 0.0041 \times \left(I_Q - 7.67 \right) \right] \end{array} \right| \le 0.1$$

$$\left| \left[0.094 - 0.0075 \times I_Q \right] + j\left[0.0356 + 0.0041 \times I_Q \right] \right| \le 0.1$$

$$\left[0.094 - 0.0075 \times I_Q \right]^2 + \left[0.0356 + 0.0041 \times I_Q \right]^2 \le 0.1^2$$

$$0.0000731 I_Q^2 - 0.001112 I_Q + 0.0101 \le 0.01$$

$$I_Q^2 - 15.34 I_Q + 1.37 \le 0$$

$$I_Q \ge 0.09 \, \text{pu}$$

If $I_Q = 0.1 \, \text{pu}$, then

$$\Delta V = \left| (8.94 - j7.57) \times (0.0041 + j0.0075) \right| = 0.099 \ \text{pu}$$

If $I_Q = 0.5 \ \text{pu}$, then

$$\Delta V = \left| (8.94 - j7.17) \times (0.0041 + j0.0075) \right| = 0.097 \ \text{pu}$$

As can be seen the improvement of the voltage at busbar B due to reactive power injection is limited.

5.3.3 Active Filtering

If the load in Figure 5.22a is a non-linear load, it absorbs harmonics from the grid. A voltage source converter connected to the load bus can act as an active filter, thus minimizing the harmonic currents drawn from the grid. As shown in Figure 5.22b, if the VSC injects equal and opposite harmonic currents, then the current drawn from the grid becomes a perfect sinusoidal. However, the operation in reality depends on the extraction of load current harmonics and the controller used to generate the injected harmonic currents.

Example 5.9

The current injected by the active filter in Figure 5.21 can be represented by $\mathbf{I_F} = G\mathbf{I_L}$ where G is the transfer function of the active filter including the detection circuit that is used to detect the harmonics in the load and the transfer function of the control circuit. For ideal operation, prove that G should be 0 for the fundamental and 1 for all the other harmonic components.

Answer

If $\mathbf{Z_s} = R + jX$, then the following equation can be written:

$\mathbf{V_S} - (\mathbf{I_L} - \mathbf{I_F})\mathbf{Z_S} = \mathbf{I_L}\mathbf{Z_L}$, where $\mathbf{Z_L}$ is the load impedance.

By substituting $\mathbf{I_F} = G\mathbf{I_L}$ into this above equation, the following was obtained:

$$\mathbf{V_S} - (\mathbf{I_L} - G\mathbf{I_L})\mathbf{Z_S} = \mathbf{I_L}\mathbf{Z_L}$$

$$\mathbf{I_L} = \frac{\mathbf{V_S}}{\left[(1-G)\mathbf{Z_S} + \mathbf{Z_L}\right]}$$

Current supply from the grid

$$\mathbf{I_S} = (\mathbf{I_L} - \mathbf{I_F}) = (1-G)\mathbf{I_L}$$

$$= (1-G)\frac{\mathbf{V_S}}{\left[(1-G)\mathbf{Z_S} + \mathbf{Z_L}\right]}$$

$$= \frac{\mathbf{V_S}}{\left[\mathbf{Z_S} + (\mathbf{Z_L}/(1-G))\right]}$$

Ideally if $G = 0$ for fundamental, then $\mathbf{I_S} = (\mathbf{V_S}/(\mathbf{Z_S} + \mathbf{Z_L}))$ and if $G = 1$ for all the other harmonics $\mathbf{I_S} \rightarrow 0$. However, in practice this cannot be achieved, and certain harmonics will be present in the current drawn from the source.

5.4 Devices for Network Support and Other Applications

5.4.1 STATCOM for Load Compensation

A part of a distribution circuit is shown in Figure 5.23. If the load connected to busbar B is a three-phase balanced load such as a large motor, then a STATCOM at the load busbar can be used to control the voltage at the point of the load.

Without the STATCOM, current in line section AB equals

$$\mathbf{I} = \frac{P - jQ}{\mathbf{V_R^*}} \tag{5.5}$$

The receiving end voltage is given by

$$\mathbf{V_S} = \mathbf{V_R} + \mathbf{I}(R + jX) \tag{5.6}$$

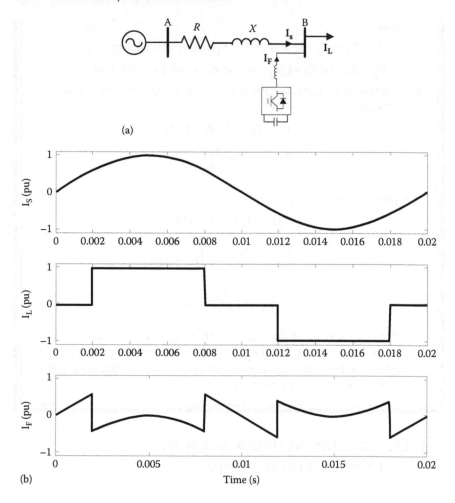

(a)

(b)

FIGURE 5.22 A shunt active compensator (a) Two bus network with the active filter. (b) Current through line AB (I_S), load (I_L) and from the active filter (I_F).

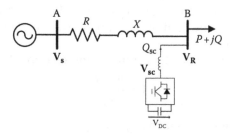

FIGURE 5.23 A STATCOM for voltage control.

From (5.5) and (5.6), the following equation is obtained:

$$\mathbf{V_S} = \mathbf{V_R} + (R + jX)\left[\frac{P - jQ}{\mathbf{V_R^*}}\right] \qquad (5.7)$$

Assuming the sending end voltage ($\mathbf{V_R}$) as the reference (i.e. $\mathbf{V_R} = V_R \angle 0°$), then Equation 5.7 can be rewritten as

$$\mathbf{V_S} = V_R + \left[\frac{RP + XQ}{V_R}\right] - j\left[\frac{XP - RQ}{V_R}\right] \qquad (5.8)$$

The right-hand side of Equation 5.8 has the receiving end voltage, a component of voltage drop in phase with V_R and a component of voltage drop perpendicular to V_R. When in-phase voltage drop is considered, the voltage drop in line AB is given by

$$|\Delta V| = \left[\frac{RP + XQ}{V_S}\right] \qquad (5.9)$$

If the STATCOM is connected as shown in Figure 5.23 and if it is supplying part of the reactive power requirement of the load, then Equation 5.9 will change to

$$|\Delta V| = \left[\frac{RP + X(Q - Q_{SC})}{V_S}\right] \qquad (5.10)$$

From Equation 5.10, it is clear that the reactive power supplied by the STATCOM reduces the voltage drop across the cable.

Example 5.10

In the network shown in Figure 5.20, calculate the reactive power that should be injected by the STATCOM to boost the voltage to 0.95 pu. Neglect the losses in line section BC.

Answer

For the motor, $P = 1.75$ pu and $Q = 1.313$ pu (since the pf is 0.8)
For the load, $P = 5.0$ pu and $Q = 2.42$ pu (since the pf is 0.9)
From Equation 5.10,

$$|\Delta V| = \left[\frac{0.0041 \times (5 + 1.75) + 0.0075 \times (2.42 + 1.313 - Q_{SC})}{1}\right] = 1 - 0.95$$

$$\therefore Q_{SC} = 0.76\,\text{pu}$$

5.4.2 STATCOM for Grid Code Compliance of Wind Farms

A typical wind farm connection is shown in Figure 5.24. Reactive power support from the wind farm varies with the active power generation (which is a function of the number of wind turbines in operation) and the voltage at the point of connection (PoC). For the following discussion, it is assumed that the voltage at the PoC is maintained at 1 pu. Without the STATCOM, the voltage drop across the wind farm cable network is given by

$$|\Delta V| = \left[\frac{RP_{\text{Wind}} + X\left(Q_{\text{Wind}} - Q_{\text{Cable}}\right)}{V_{\text{S}}} \right] \tag{5.11}$$

where R and X are equivalent resistance and reactance.

When the power generated by the wind farm is zero ($P_{\text{Wind}} = 0$) due to the reactive power supplied by the cable network (Q_{Cable}), the voltage at the PoC tries to increase beyond 1 pu ($|\Delta V| < 0$). The only way to maintain the voltage at the PoC at 1 pu is by absorbing a part of the reactive power supplied by the cable (Q_{Cable}) using the converters in the wind turbines. However, as the wind farm generates power, the reactive power that can be absorbed by the turbines decreases (their converters are rated just to operate at the rated power factor when they are generating 100% active power). This is shown

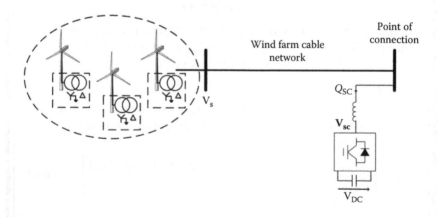

FIGURE 5.24 A wind farm connection.

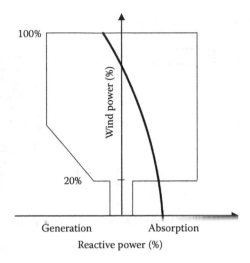

FIGURE 5.25 Reactive power characteristic of a wind farm (thick line) and grid code requirement (thin line).

in Figure 5.25. In the figure, grid code requirements of the United Kingdom and Ireland are shown. It is obvious that in order to fulfil the grid code requirement for a given active power generation, a reactive power compensator is required. For this, a STATCOM, together with switched capacitors and reactors, is employed in present-day wind farms.

5.4.3 Soft Open Points for Network Support

Soft open points (SOPs) are power electronic devices installed in place of previously normally open points in electrical power distribution networks. They can provide active power flow control, reactive power compensation and voltage regulation under normal network operating conditions, as well as fast fault isolation and supply restoration under abnormal conditions. An SOP connected at the end points of two feeders of a distribution network is shown in Figure 5.26.

SOPs operate in two different modes: normal and restoration operation. The normal operation mode usually uses a current control and provides independent control of real and reactive power. A supply restoration mode usually uses a voltage control and enables power supply to isolate loads due to network faults.

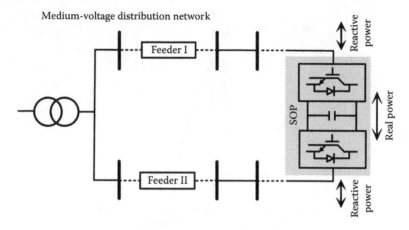

FIGURE 5.26 A distribution network with an SOP connected at the ends of two feeders.

5.4.3.1 Normal Operation Mode

A dual closed-loop current-controlled strategy [12] is usually used to operate the SOP for the control of the feeder power flow under normal network operating conditions. Such a current-controlled strategy is advantageous because (a) it provides decoupled control of active and reactive power components and (b) it inherently limits the VSC current during network faults.

Figure 5.27 presents the overall control structure. The outer power control loop, the inner current control loop and the phase-locked loop (PLL) are the three main components. In the outer power control loop (Figure 5.27a), one of the VSCs operates in the P–Q control scheme where the active and reactive power errors are transformed into the reference d–q current components, i_d^* and i_q^*, through the PI controllers. The superscript asterisk denotes reference value. The other VSC operates in a V_{DC}–Q control scheme maintaining a constant DC-side voltage for a stable and balanced active power flow through the DC link. Dynamic limiters for i_d^* and i_q^* are inserted to enable overcurrent limiting during network faults and disturbances. In the inner current control loop (Figure 5.27b), the reference VSC d–q voltage, V_{dm} and V_{qm} are determined through the PI controllers considering the d–q current errors. The voltage feedforward and current feedback compensations are used to get a good dynamic response [13]. After transforming V_{dm} and V_{qm} into the VSC terminal voltage by Park's transformation [14], the gate signals for the IGBTs are obtained through the PWM. The PLL is important for the connection of VSCs to the AC network in order to synchronize the output VSC voltage

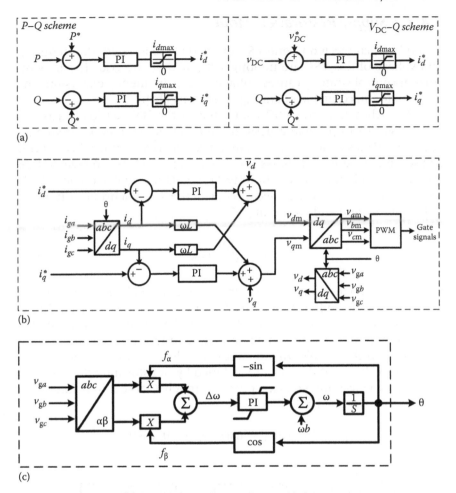

FIGURE 5.27 Control block diagram of the SOP for normal operation mode: (a) power control loop, (b) current control loop and (c) PLL.

with the AC network voltage. A PLL control topology based on the pq theory was used [15], as shown in Figure 5.27c. By using the sum of the products of the feedback signals, f_α and f_β, and input $\alpha-\beta$ voltages transformed through Clark's transformation [14], the variation of the angular frequency $\Delta\omega$ is calculated as

$$\Delta\omega = V_\alpha \cdot f_\beta + V_\beta \cdot f_\alpha \qquad (5.12)$$

The PLL output angle θ with a frequency $f = 2\pi \cdot \omega$ is then obtained using a PI-controller, a feedback compensation of the base angular frequency ω_b and an integrator.

5.4.3.2 Supply Restoration Mode

When loads connected to one VSC of an SOP are isolated, the frequency and voltage of this VSC are no longer dictated by the grid. The previous current control strategy causes voltage and/or frequency excursions that can lead to unacceptable operating conditions [16].

In such a case, the VSC connected to the isolated loads acts as a voltage source to provide a desired load voltage with stable frequency. The other VSC still acts as a current source operating in the $V_{DC}-Q$ control scheme. Figure 5.28 shows the block diagram of the voltage and frequency control strategy for the interface VSC. For voltage control, the VSC output voltage is controlled directly in the d–q synchronous frame by holding V_q^* to zero and controlling V_d^* as

$$V_d = \sqrt{\frac{2}{3}} \cdot V_{rms}^* \qquad (5.13)$$

where V_{rms}^* is the desired nominal line-to-line rms voltage of the isolated loads. $\sqrt{2/3}$ is included because Park's transformation is based on the peak value of the phase voltage. The output VSC voltage is regulated by a closed-loop control and generated through the PWM scheme. For frequency control, a stable voltage frequency is generated using the PLL. The input voltages of the PLL are assigned by transforming the same d–q reference voltages for voltage control through Clarke's transformation. Thus, $\Delta\omega$ is calculated as

$$\Delta\omega = V_d \cdot \cos\theta \cdot (-\sin\theta) + V_d \cdot \sin\theta \cdot \cos\theta = 0 \qquad (5.14)$$

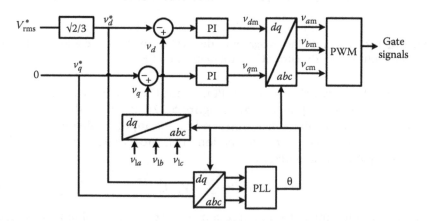

FIGURE 5.28 Control block diagram of the interface VSC for the supply restoration control mode.

Since $\Delta\omega$ remains zero, the phase angle θ with a constant frequency $f = 2\pi \cdot \omega_b$ is generated by integrating only the base angular frequency ω_b, as shown in Figure 5.27.

5.4.4 Fault Current Limiter for Network Support

With the connection of DERs, the fault level at some busbars may go increase beyond the breaking capacity of the switch gear. Under such circumstances, it may be economic to use a fault current limiter (FCL) rather than replacing the switch gear. The simplest form of FCL is a current limiting reactor (CLR), as shown in Figure 5.29. In addition to the connections shown in Figure 5.29, a CLR may be connected to an incoming feeder or an outgoing feeder. One of the problems with a CLR is the resistive component of the reactor; this series-connected component continuously produces heat. Further, it produces a voltage drop under normal operating conditions. One solution to this is to use a thyristor-controlled FCL, as shown in Figure 5.30. Here, under normal operations, thyristors are

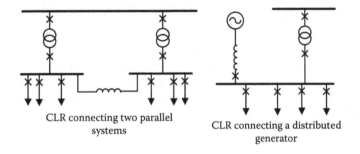

CLR connecting two parallel systems

CLR connecting a distributed generator

FIGURE 5.29 Possible connection of CLR.

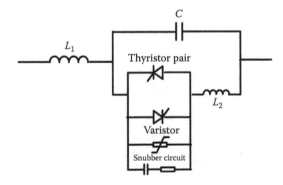

FIGURE 5.30 Thyristor-controlled FLC.

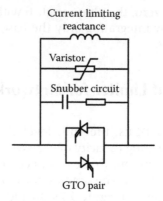

FIGURE 5.31 GTO-connected FLC.

blocked and current passes through a reactor-capacitor branch. The reactor, L_1, has a much smaller reactance than that of a CLR, thus heat losses are much less. The values of L_1 and C are selected such that at 50 Hz, the resultant impedance is only the resistive component of the reactor, thus minimizing the voltage drop. During a fault, the thyristor pair is triggered; thus, the fault current passes through L_1 and L_2, thus limiting the fault current.

Another power electronics–based FCL is a GTO-based circuit, as shown in Figure 5.31. The back-to-back gate turn off (GTO) thyristors are the main conductive path during normal operation. When a fault is detected, the GTOs are switched off and the current is diverted to the parallel reactance. This limits the fault current.

5.4.5 Solid-State Transformer for Network Support

As shown in Figure 5.32, a solid-state transformer (SST) typically includes a medium-voltage AC–DC power conversion stage to generate

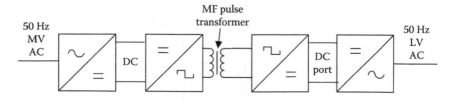

FIGURE 5.32 A solid-state transformer.

a medium voltage DC bus, a medium-frequency (0.5–1 kHz) DC–DC converter stage to produce a regulated low-voltage DC bus and a DC–AC converter stage to produce a regulated low-voltage AC bus [15,16]. As discussed in [16], in order to achieve the required voltage, a modular design based on a number of H-bridge modules connected in series at the input side is considered. These converters are switched using an interleaved carrier-based PWM technique. Each PWM carrier is phase-shifted by 360°/N (where N is the series-connected module). The PWM for each cell is generated by comparing a reference signal with a corresponding PWM carrier, as shown in Figure 5.3. An SST can be considered as a three-port energy router which integrates the distribution system, a residential AC system, and a DC system. In order to improve system efficiency, the DC type sources and energy storage can be connected to the DC port.

☐ Questions

1. A PV module has a characteristic shown in Figure 5.7. A PV array is formed by connecting 15 PV modules in series. The array is connected to a central DC–DC full bridge converter and an inverter. If the inverter maintain the dc link voltage at 325 V, what should be the duty ratio of the DC–DC converter to extract maximum power at irradiance of 1000 and 500 W/m²?

2. Figure 5.33 shows the reactive power demanded by the UK grid code and the active power vs reactive power for a 100 MW wind farm under different voltage at the PoC. Design a suitable reactive power compensation scheme to fulfil the grid code requirements.

3. In the distribution network shown in Figure 5.20, a DVR is connected to line AB at end B. Answer the following:

   a. What is the voltage at busbar B, if the motor is not in operation?

   b. When the motor is connected, it draws three times the rated current at 0.8 power factor lagging for 50 ms. What is the voltage during the sag?

   c. Calculate the voltage required to be injected by the DVR if (i) active power compensation: pre sag and (ii) active power compensation: in-phase is employed by the DVR.

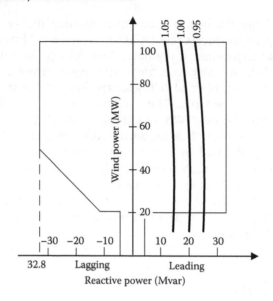

FIGURE 5.33 Figure for Question 2.

☐ References

1. H. W. Van der Broeck, H. C. Skudelny and G. V. Stanke, Analysis and realization of a pulsewidth modulator based on voltage space vectors, *IEEE Transactions on Industry Applications*, 24(1), 142–150, January/February 1988.
2. P. G. Handley and J. T. Boys, Space vector modulation – An engineering review, *IEE Fourth International Conference on Power Electronics and Variable Speed Drives (PEVD)*, London, UK, 17–19 July 1990, pp. 87–91.
3. E. Roman, R. Alonso, P. Ibanez, S. Elorduizapatarietxe and D. Goitia, Intelligent PV module for grid-connected PV systems, *IEEE Transactions on Industrial Electronics*, 53(4), 1066–1073, June 2006.
4. N. Mohan and T. M. Undeland, *Power Electronics: Converters, Applications, and Design*. New Delhi, India: Wiley: 2007, pp. 178–183.
5. R. W. Erickson and D. Maksimovic, *Fundamentals of Power Electronics*. Kluwer Academic Publisher, New York, 2004, pp. 38–53.
6. K. M. Rahman and M. Ehsani, Performance analysis of electric motor drives for electric and hybrid electric vehicle applications, *IEEE Power Electronics for Transportation*, 1996, 49–56.
7. E. Schaltz, Electrical vehicle design and modeling, electric vehicles – Modelling and simulations, 2011. Available from: http://www.intechopen.com/books/electric-vehicles-modelling-and-simulations/electrical-vehicle-design-and-modeling, Accessed on 05/09/2014.

8. N. Abi-Samra, C. Neft, A. Sundaram and W. Malcolm, The distribution system Dynamic Voltage Restorer and its application at industrial facilities with sensitive loads, *Power Conversion and Intelligent Motion Power System World Conference*, Nurnbery, Germany, 20–22 June 1995, pp. 1–15.
9. G. T. Heydt, W. Tan, T. LaRose and M. Negley, Simulation and analysis of series voltage boost technology for power quality enhancement, *IEEE Transaction on Power Delivery*, 13(4), 1335–1341, October 1998.
10. C. S. Chang, S. W. Yang and Y. S. Ho, Simulation and analysis of Series Voltage Restorers (SVR) for voltage sag relief, *IEEE Power Engineering Society Winter Meeting*, Singapore, 23–27 January 2000, pp. 2476–2481.
11. S. S. Choi, B. H. Li and D. M. Vilathgamuwa, Dynamic voltage restoration with minimum energy injection, *IEEE Transactions on Power Systems*, 15(1), 51–57, February 2000.
12. R. S. Weissbach, G. G. Karady and R. G. Farmer, Dynamic voltage compensation on distribution feeders using flywheel energy storage, *IEEE Transactions on Power Delivery*, 14(2), 465–471, April 1999.
13. N. Chaudhuri, R. Majumder, B. Chaudhuri and P. Jiuping, Stability analysis of VSC MTDC grids connected to multimachine AC systems, *IEEE Transactions on Power Delivery*, 26, 2774–2784, 2011.
14. J. Espi Huerta, J. Castello-Moreno, J. Fischer and R. Garcia-Gil, A synchronous reference frame robust predictive current control for three-phase grid-connected inverters, *IEEE Transactions on Industrial Electronics*, 57,954–962, 2010.
15. I. Balaguer, L. Qin, Y. Shuitao, U. Supatti and P. Zheng, Control for grid-connected and intentional islanding operations of distributed power generation, *IEEE Transactions on Industrial Electronics*, 58,147–157, 2011.
16. L. Rolim, D. da Costa and M. Aredes, Analysis and software implementation of a robust synchronizing PLL circuit based on the pq theory, *IEEE Transactions on Industrial Electronics*, 53, 1919–1926, 2006.
17. S. Bifaretti, P. Zanchetta, A. Watson, L. Tarisciotti and J. Clare, Advanced power electronic conversion and control system for universal and flexible power management, *IEEE Transactions on Smart Grid*, 2(2), 231–243, June 2011.

6 CHAPTER

Operation Simulation and Analysis

6.1 Introduction

The development of a smart distribution system needs the support of more specialized simulation and analysis methods [1,2]. For component modelling, flexible approaches should be adopted in response to the emerging loads, distributed generators (DGs) and other new devices. Meanwhile, given that there can be more smart terminals with new control functions, the mathematical models of different controllers need to be established. For simulation and analysis, DGs have a large impact on the existing power flow programs and will bring a series of challenges to short-circuit current calculation, reliability analysis and so on. Besides that, considering the stochastic fluctuation of DGs, simulation methods not only need to analyse the situation for one certain time point, but also that for a series of time points to show the operation state of a smart distribution system.

Therefore, traditional simulation methods need to be extended and improved. This chapter introduces the key simulation and analysis methods, including steady-state modelling, power flow analysis, sequential power flow analysis and fault calculation.

6.2 Distribution Network Components and Their Modelling

For a given network structure and operation condition, power flow analysis calculates nodal voltages (amplitudes and phase angles), power distribution and losses of the entire network. Power flow analysis is important for quantifying the reliability and economics of the distribution system [3].

As various DGs and energy storage systems (ESSs) are integrated into a smart distribution network, the power flow analysis of such a network is more challenging than a conventional distribution network. First, the components and their modelling of a distribution network need to be studied.

6.2.1 Distribution Line

The distribution line model is shown in Figure 6.1. In the π type equivalent circuit of a three-phase line, Z_l represents the series impedance matrix and Y_l represents the shunt admittance matrix. Both Z_l and Y_l are $n \times n$ complex matrices, where n is the number of line phases. For a single-phase, two-phase and three-phase line, n takes the values of 1, 2 and 3.

The series impedance matrix is

$$\mathbf{Z}_l = \begin{bmatrix} z_{aa} & z_{ab} & z_{ac} \\ z_{ba} & z_{bb} & z_{bc} \\ z_{ca} & z_{cb} & z_{cc} \end{bmatrix} \tag{6.1}$$

where

z_{aa}, z_{bb}, z_{cc} are the self-impedances of AC lines

$z_{ab}, z_{ba}, z_{AC}, z_{ca}, z_{bc}, z_{cb}$ are the mutual impedances

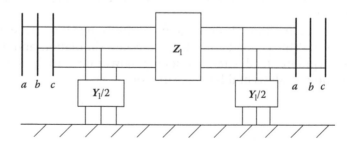

FIGURE 6.1 Three-phase distribution line model.

The shunt admittance matrix is

$$\frac{Y_1}{2} = \frac{1}{2} \times \begin{bmatrix} y_{aa} & y_{ab} & y_{ac} \\ y_{ba} & y_{bb} & y_{bc} \\ y_{ca} & y_{cb} & y_{cc} \end{bmatrix} \quad (6.2)$$

where
  y_{aa}, y_{bb}, y_{cc} are the self-admittances of AC lines
  $y_{ab}, y_{ba}, y_{AC}, y_{ca}, y_{bc}, y_{cb}$ are the mutual admittances

From Equations 6.1 and 6.2, the admittance matrix Y_L of the distribution line model is represented as

$$Y_L = \begin{bmatrix} Z_1^{-1} + \dfrac{1}{2}Y_1 & -Z_1^{-1} \\ -Z_1^{-1} & Z_1^{-1} + \dfrac{1}{2}Y_1 \end{bmatrix} \quad (6.3)$$

Sometimes the influence of shunt admittances is neglected, thus the line model is simplified to the admittance matrix shown in Equation 6.4.

$$Y_L = \begin{bmatrix} Z_1^{-1} & -Z_1^{-1} \\ -Z_1^{-1} & Z_1^{-1} \end{bmatrix} \quad (6.4)$$

Example 6.1

The impedances per unit length of a 2 km long 10 kV overhead line are

$$z_{aa} = z_{bb} = z_{cc} = (0.550 + j0.727) \ \Omega/\text{km}$$
$$z_{ab} = z_{ba} = z_{ac} = z_{ca} = z_{bc} = z_{cb} = (0.220 + j0.291) \ \Omega/\text{km}$$

Obtain the series impedance matrix.

Answer
Total self-impedances are $z_{aa} = z_{bb} = z_{cc} = (0.550 + j0.727) \times 2 =$ (1.1 + j1.454) Ω
Total mutual impedances are

$$z_{ab} = z_{ba} = z_{ac} = z_{ca} = z_{bc} = z_{cb} = (0.220 + j0.291) \times 2 \ \Omega$$
$$= (0.44 + j0.582) \ \Omega$$

Substituting self-impedances and mutual impedances into Equation 6.1, the series impedance matrix is obtained as follows:

$$\mathbf{Z}_1 = \begin{bmatrix} 1.100 + j1.454 & 0.440 + j0.582 & 0.440 + j0.582 \\ 0.440 + j0.582 & 1.100 + j1.454 & 0.440 + j0.582 \\ 0.440 + j0.582 & 0.440 + j0.582 & 1.100 + j1.454 \end{bmatrix} \ \Omega$$

6.2.2 Distribution Transformer

In order to obtain the equivalent circuit of a transformer, the resistance and leakage reactance of the secondary winding is transformed to the primary side. Merged with the resistance and reactance of the primary winding, the equivalent resistance and reactance are represented as R_T and X_T. Besides, G_T and B_T are the equivalent conductance and susceptance of the magnetizing branch. These four equivalent parameters can be calculated by nameplate ratings – short-circuit loss ΔP_k, short circuit voltage percentage $V_k\%$, no-load loss ΔP_0 and no-load current percentage $I_0\%$. Since G_T and B_T are relatively small, they can be ignored in general [4].

When modelling a transformer, its connection arrangement should be considered. A delta-grounded wye step-down (D/Y_g) transformer shown in Figure 6.2a is used as an example to illustrate the modelling method. The primary referred equivalent circuit is shown in Figure 6.2b.

The leakage admittance of the primary winding is given as $y_T = 1/(R_T + jX_T)$. The relationship between node voltage and node injection current in per unit can be described using the following equations, where transformation ratio $k = 1$. A letter with a dot placed over it denotes a phasor.

$$\begin{cases} \dot{I}_a^p = \dot{I}_a^r - \dot{I}_c^r \\ \dot{I}_b^p = \dot{I}_b^r - \dot{I}_a^r \\ \dot{I}_c^p = \dot{I}_c^r - \dot{I}_b^r \end{cases} \tag{6.5}$$

$$\dot{I}_a^r = -\dot{I}_a^s = y_T \left[\left(\dot{V}_a^p - \dot{V}_b^p \right) - \dot{V}_a^s \right]$$

$$\dot{I}_b^r = -\dot{I}_b^s = y_T \left[\left(\dot{V}_b^p - \dot{V}_c^p \right) - \dot{V}_b^s \right] \tag{6.6}$$

$$\dot{I}_c^r = -\dot{I}_c^s = y_T \left[\left(\dot{V}_c^p - \dot{V}_a^p \right) - \dot{V}_c^s \right]$$

where
   $\dot{I}_a^p$, $\dot{I}_b^p$ and $\dot{I}_c^p$ are the primary node current of phase a, b, c
   $\dot{I}_a^s$, $\dot{I}_b^s$ and $\dot{I}_c^s$ are the secondary node current of phase a, b, c
   $\dot{I}_a^r$, $\dot{I}_b^r$ and $\dot{I}_c^r$ are the primary node current of phase a, b, c after delta-wye transformation

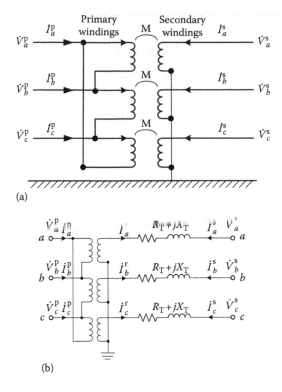

FIGURE 6.2 Delta-grounded wye connection: (a) winding connection and (b) the per unit equivalent circuit.

Substitute Equation 6.6 into Equation 6.5 and the transformer's node admittance matrix has the relationship as

$$
\dot{I}_n =
\begin{bmatrix}
\dot{I}_a^p \\
\dot{I}_b^p \\
\dot{I}_c^p \\
\dot{I}_a^s \\
\dot{I}_b^s \\
\dot{I}_c^s
\end{bmatrix}
=
\begin{bmatrix}
y_T \begin{bmatrix} 2 & -1 & -1 \\ -1 & 2 & -1 \\ -1 & -1 & 2 \end{bmatrix} & -y_T \begin{bmatrix} 1 & 0 & -1 \\ -1 & 1 & 0 \\ 0 & -1 & 1 \end{bmatrix} \\
-y_T \begin{bmatrix} 1 & -1 & 0 \\ 0 & 1 & -1 \\ -1 & 0 & 1 \end{bmatrix} & y_T \begin{bmatrix} 1 & 0 & 0 \\ 0 & 1 & 0 \\ 0 & 0 & 1 \end{bmatrix}
\end{bmatrix}
\begin{bmatrix}
\dot{V}_a^p \\
\dot{V}_b^p \\
\dot{V}_c^p \\
\dot{V}_a^s \\
\dot{V}_b^s \\
\dot{V}_c^s
\end{bmatrix}
= Y_T \dot{V}_n
$$

$$(6.7)$$

where
    Y_T is transformer's node admittance matrix
    $\dot{I}_n = \begin{bmatrix} \dot{I}_a^p & \dot{I}_b^p & \dot{I}_c^p & \dot{I}_a^s & \dot{I}_b^s & \dot{I}_c^s \end{bmatrix}^T$ is the node current vector
    $\dot{V}_n = \begin{bmatrix} \dot{V}_a^p & \dot{V}_b^p & \dot{V}_c^p & \dot{V}_a^s & \dot{V}_b^s & \dot{V}_c^s \end{bmatrix}^T$ is the node voltage vector

Example 6.2

A two-winding 20,000 kVA, 110/10.5 kV transformer of delta-grounded wye connection (D/Y_g) has the following parameters:

$\Delta P_k = 88.280$ kW, $V_k\% = 10.76$, $\Delta P_0 = 17.110$ kW, $I_0\% = 0.22$.

Please determine the node admittance matrix of this transformer.

Answer

Using the nameplate ratings, the resistance, reactance, conductance and susceptance are calculated as

$$R_T = \frac{\Delta P_0 V_N^2}{S_N^2} \times 10^3 = \frac{88.280 \times 110^2}{20,000^2} \times 10^3 = 2.670 \ \Omega$$

$$X_T = \frac{V_0\%}{100} \frac{V_N^2}{S_N} \times 10^3 = \frac{10.76 \times 110^2}{100 \times 20,000} \times 10^3 = 65.098 \ \Omega$$

$$G_T = \frac{\Delta P_0}{V_N^2} \times 10^{-3} = \frac{17.110}{110^2} \times 10^{-3} = 1.414 \times 10^{-6} \ S$$

$$B_T = \frac{I_0\%}{100} \frac{S_N}{V_N^2} \times 10^{-3} = \frac{0.22 \times 20,000}{100 \times 110^2} \times 10^{-3} = 3.636 \times 10^{-6} \ S$$

It is seen that G_T and B_T are relatively small, so they are ignored. Then, the short-circuit admittance of the primary winding is

$$y_T = \frac{1}{R_T + jX_T} = \frac{1}{2.670 + j65.098} = (0.063 - j1.534) \times 10^{-2} \ S$$

Substitute y_T into Equation 6.7:

$$Y_T = \left((0.063 - j1.534) \times 10^{-2} \right) \times \begin{bmatrix} \begin{bmatrix} 2 & -1 & -1 \\ -1 & 2 & -1 \\ -1 & -1 & 2 \end{bmatrix} & - \begin{bmatrix} 1 & 0 & -1 \\ -1 & 1 & 0 \\ 0 & -1 & 1 \end{bmatrix} \\ - \begin{bmatrix} 1 & -1 & 0 \\ 0 & 1 & -1 \\ -1 & 0 & 1 \end{bmatrix} & \begin{bmatrix} 1 & 0 & 0 \\ 0 & 1 & 0 \\ 0 & 0 & 1 \end{bmatrix} \end{bmatrix}$$

6.2.3 Distribution Load

The loads on a distribution system can be modelled as wye- or delta-connected [5]. The distribution load can be three-phase, two-phase (e.g. phases a and c), or single-phase (e.g. phase b) with any degree of unbalance, as shown in Figure 6.3.

The load can be divided into three types: constant real and reactive power (constant PQ), constant current and constant impedance, as shown in Table 6.1.

Example 6.3

A load rated at 10 kV, 2.35 MW is connected to a 10 kV distribution system. If the current load voltage is 9.92 kV, calculate the actual real power consumed by the load under (a) constant PQ, (b) constant current and (c) constant impedance representations.

Answer

a. Constant PQ:
   The actual power is $P = P_N = 2.35$ MW.
b. Constant current:
   The rated current of the load is

$$I_N = \frac{P_N}{V_N} = \frac{2.35 \times 10^6}{10 \times 10^3} = 235\,\text{A}$$

   The actual power is

$$P = VI_N = 9.92 \times 10^3 \times 235 = 2.331\,\text{MW}$$

c. Constant impedance:
   The impedance of the load is

$$R_N = \frac{V_N^2}{P_N} = \frac{\left(10 \times 10^3\right)^2}{2.35 \times 10^6} = 42.553\,\Omega$$

   The actual power is

$$P = \frac{V^2}{R_N} = \frac{\left(9.92 \times 10^3\right)^2}{42.553} = 2.313\,\text{MW}$$

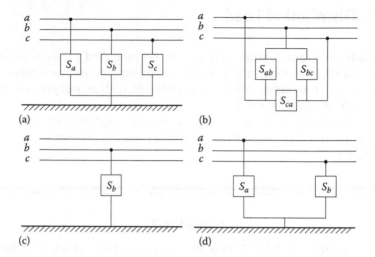

FIGURE 6.3 Load model: (a) grounded wye-connected three-phase load, (b) ungrounded delta-connected three-phase load, (c) grounded single-phase load and (d) grounded two-phase load.

TABLE 6.1 Different Types of Load

Type	Description
Constant PQ	The power consumption of loads does not vary with load voltages.
Constant current	The power consumption of loads changes linearly with load voltages.
Constant impedance	The power consumption of loads changes with load voltages in a quadratic function relationship.

6.2.4 Shunt Capacitor

In distribution networks, shunt capacitor banks are widely used for reactive power compensation. When the grid voltage is below the lower limit, capacitors are switched on in order to improve the voltage level. Conversely, when the grid voltage is above the upper limit, capacitors are switched off. Assuming that there are several capacitor banks in the system with different capacitance, the admittance matrix of each capacitor

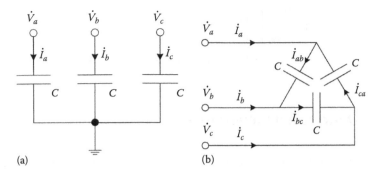

FIGURE 6.4 Shunt capacitor connections: (a) grounded wye connection and (b) ungrounded delta connection.

can be calculated separately. The total admittance matrix of the whole shunt capacitors can be obtained by finding the sum of individual matrix.

Shunt capacitor banks can be connected in wye configuration or delta configuration as shown in Figure 6.4.

Assume that a three-phase capacitor is wye connected and the admittance in each phase is y_c, then the admittance matrix of the capacitor is given by

$$\begin{bmatrix} \dot{I}_a \\ \dot{I}_b \\ \dot{I}_c \end{bmatrix} = \begin{bmatrix} y_c & 0 & 0 \\ 0 & y_c & 0 \\ 0 & 0 & y_c \end{bmatrix} \begin{bmatrix} \dot{V}_a \\ \dot{V}_b \\ \dot{V}_c \end{bmatrix} = Y_Y \begin{bmatrix} \dot{V}_a \\ \dot{V}_b \\ \dot{V}_c \end{bmatrix} \tag{6.8}$$

For the delta connection, the relationship between node voltage and current can be described as Equations 6.9 through 6.11. Substitute Equations 6.10 and 6.11 into Equation 6.9 and the admittance matrix can be transformed into Equation 6.12.

$$\begin{bmatrix} \dot{I}_a \\ \dot{I}_b \\ \dot{I}_c \end{bmatrix} = \begin{bmatrix} 1 & 0 & -1 \\ -1 & 1 & 0 \\ 0 & -1 & 1 \end{bmatrix} \begin{bmatrix} \dot{I}_{ab} \\ \dot{I}_{bc} \\ \dot{I}_{ca} \end{bmatrix} \tag{6.9}$$

$$\begin{bmatrix} \dot{I}_{ab} \\ \dot{I}_{bc} \\ \dot{I}_{ca} \end{bmatrix} = \begin{bmatrix} y_c & 0 & 0 \\ 0 & y_c & 0 \\ 0 & 0 & y_c \end{bmatrix} \begin{bmatrix} \dot{V}_a - \dot{V}_b \\ \dot{V}_b - \dot{V}_c \\ \dot{V}_c - \dot{V}_a \end{bmatrix} \tag{6.10}$$

$$\begin{bmatrix} \dot{V}_a - \dot{V}_b \\ \dot{V}_b - \dot{V}_c \\ \dot{V}_c - \dot{V}_a \end{bmatrix} = \begin{bmatrix} 1 & -1 & 0 \\ 0 & 1 & -1 \\ -1 & 0 & 1 \end{bmatrix} \begin{bmatrix} \dot{V}_a \\ \dot{V}_b \\ \dot{V}_c \end{bmatrix} \tag{6.11}$$

$$\begin{bmatrix} \dot{I}_a \\ \dot{I}_b \\ \dot{I}_c \end{bmatrix} = \begin{bmatrix} 2y_c & -y_c & -y_c \\ -y_c & 2y_c & -y_c \\ -y_c & -y_c & 2y_c \end{bmatrix} \begin{bmatrix} \dot{V}_a \\ \dot{V}_b \\ \dot{V}_c \end{bmatrix} = Y_D \begin{bmatrix} \dot{V}_a \\ \dot{V}_b \\ \dot{V}_c \end{bmatrix} \tag{6.12}$$

6.2.5 Distributed Generators

There are many kinds of DGs, and they can be divided into dispatch-able DGs, such as micro gas turbines (MT) and fuel cells (FC), and non-dispatchable DGs, such as wind turbines (WT) and photovoltaic (PV) systems. Non-dispatchable DGs mainly generate electricity through the utilization of natural resources, thus the power output is influenced by the environment and climate such as wind speed, solar irradiance and ambient temperature. Such DGs are environmentally friendly and do not consume any fuel, so it is usually necessary to make the most use of the electricity they generate. In power flow analysis, the power outputs of these DGs can be calculated considering the environmental factors.

Taking PV as an example, as a DC power supply device, the PV panel is integrated into the grid through a power electronic interface, which inverts DC electricity to AC electricity [6]. Figure 6.5 illustrates the entire structure of a PV model.

In Figure 6.5, P_{mpp} represents the maximum output power of a PV panel under the IEC standard condition (the solar irradiance is 1000 W/m², the operating temperature is 298 K, and the air mass is 1.5); T and I_{rr} are temperature and irradiance.

The temperature and irradiance influence the PV output power. P_{mpp} needs to be modified when there is discrepancy between the actual environmental factors and the standard condition. In steady-state analy-sis, it is considered that the PV output power is approximately propor-tional to irradiance. Its relationship with temperature can be described by

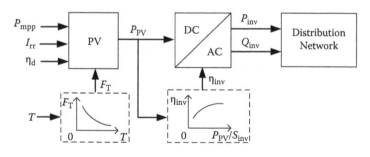

FIGURE 6.5 The PV model.

the F_T–T curve shown in Figure 6.5. F_T is a correction factor of the output power, which is equal to standard conditions. It must be noted that the surface of a PV panel can be stained by dust, rainwater, etc. Considering the stain coefficient η_d, which reflects the effect of environment contamination on the PV panel surface, the output power of PV panel P_{PV} becomes

$$P_{PV} = \eta_d I_{rr} P_{mpp} F_T \qquad (6.13)$$

The efficiency of the PV inverter can be assumed as a constant. Considering the efficiency η_{inv}, the output power of the PV inverter P_{inv} becomes

$$P_{inv} = \eta_{inv} P_{i} \qquad (6.14)$$

In general, the power factor of the PV inverter is unity, namely its reactive power is zero. In practice, the reactive power can be adjusted in response to optimal operation of distribution systems. In steady-state analysis, the reactive power Q_{spec} is given directly or calculated by the power factor *pf*.

Example 6.4

A PV system is rated $550\ kW_p$ under the IEC standard conditions. The environmental factors and other parameters of the PV system are

> Irradiance $I_{rr} = 0.95\ kW/m^2$
> Temperature $T = 298\ K$, and the corresponding correction
>   factor of the output power $F_T = 1$
> The stain coefficient of the PV panel $\eta_d = 0.98$
> The efficiency of the PV inverter $\eta_{inv} = 0.97$

Calculate the output power of this PV system.

Answer

The power output of the PV system is calculated by using Equations 6.13 and 6.14:

$$P_{PV} = \eta_{inv}(\eta_d I_{rr} P_{mpp} F_T) = 0.97 \times 0.98 \times 0.95 \times 550 \times 1 = 496.69\ kW$$

In power flow analysis, one method of simplifying these DGs is to treat them as 'negative loads' and to process them as constant *PQ* loads. But considering the characteristics of various DGs, simplifying them as constant real and reactive loads is not always reasonable. To accurately reflect characteristics of DGs, they can be modelled as *PV* nodes, *PI* nodes, *PQ* nodes or other node types, which are presented as follows. The selection is based on the generator type, the grid interface and the control strategy.

6.2.5.1 Power Electronic Devices

Among DGs, micro gas turbines and some wind turbines are integrated into the grid through two back-to-back AC-DC converters, while PV cells and fuel cells through DC-AC inverters since the electricity they generate is DC. The majority of DGs are integrated into the grid through voltage source converters (VSC) and they can be modelled as *PV* nodes, whose real power and voltage magnitude can be controlled. Current source converters (CSC) are occasionally used as the grid interface and they are modelled as *PI* nodes, whose real power and injection current magnitude are constant.

6.2.5.2 Synchronous Generators

Synchronous generators are employed in some DG applications. As their voltage regulators can keep the terminal voltage constant, they are treated as *PV* nodes. In some cases, synchronous generators maintain the power and power factor at pre-determined values. These generators are treated as *PQ* nodes.

6.2.5.3 Asynchronous Generators

Some small wind turbines and small hydro turbines employ asynchronous generators. Asynchronous generators establish magnetic fields depending on the reactive power provided by the grid, so they do not have the ability of voltage regulation. Asynchronous generators absorb reactive power from the grid when they generate real power. Therefore,

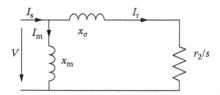

FIGURE 6.6 The equivalent circuit of an asynchronous generator.

asynchronous generators are considered as $PQ(V)$ nodes, where $Q(V)$ represents influence of V on Q. This is shown below.

Figure 6.6 gives the equivalent circuit of an asynchronous generator. In the figure, s is the slip, I_s is the stator current, I_r is the rotor current, I_m is the magnetizing current, r_2 is the rotor resistance, x_m is the magnetizing reactance and x_σ is the leakage reactance.

The output power P_e and power factor angle φ are derived from Figure 6.6:

$$P_e = \frac{s r_2 V^2}{s^2 x_\sigma^2 + r_2^2} \qquad (6.15)$$

$$\tan \varphi = \frac{r_2^2 + x_\sigma \left(x_m + x_\sigma \right) s^2}{r_2 x_m s} \qquad (6.16)$$

The slip s is deduced from Equation 6.17:

$$s = \frac{r_2 \left(V^2 - \sqrt{V^4 - 4 x_\sigma^2 P_e^2} \right)}{2 P_e x_\sigma^2} \qquad (6.17)$$

For a given real power, the reactive power Q is calculated as

$$Q = P_e \tan \varphi = \frac{r_2^2 + x_\sigma \left(x_m + x_\sigma \right) s^2}{r_2 x_m s} P_e \qquad (6.18)$$

From Equation 6.18, it can be seen that Q is a function of slip s. As slip is a function of voltage V (see Equation 6.17), the reactive power absorbed by an asynchronous generator varies with the terminal voltage of the generator.

Example 6.5

The following parameters are given for a three-phase asynchronous generator:

Real power $P_e = 3.737$ kW
Terminal voltage $V = 380$ V
Rotor resistance $r_2 = 1.047\ \Omega$
Magnetizing reactance $x_m = 82.600\ \Omega$
Leakage reactance $x_\sigma = 6.830\ \Omega$

Calculate the output reactive power of this generator. And what is the output reactive power of this generator if the terminal voltage is 400 V?

Answer

First, compute the slip based on Equation 6.17:

$$s = \frac{1.047 \times \left(380^2 - \sqrt{380^4 - 4 \times 6.830^2 \times 3737^2}\right)}{2 \times 3737 \times 6.830^2} = 0.028$$

Then, according to Equation 6.16, the tangent of the power factor angle is

$$\tan \varphi = \frac{1.047^2 + 6.83 \times (82.600 + 6.83) \times 0.028^2}{1.047 \times 82.6 \times 0.028} = 0.650$$

Substitute P_e and $\tan \varphi$ into Equation 6.18:

$$Q = P_e \tan \varphi = 3.737 \times 0.650 = 2.429 \text{ kVar}$$

If the terminal voltage is 400 V, the corresponding parameter is as follows:

$$s = \frac{1.047 \times \left(400^2 - \sqrt{400^4 - 4 \times 6.830^2 \times 3737^2}\right)}{2 \times 3737 \times 6.830^2} = 0.025$$

$$\tan \varphi = \frac{1.047^2 + 6.83 \times (82.600 + 6.83) \times 0.025^2}{1.047 \times 82.6 \times 0.025} = 0.684$$

$$Q = P_e \tan \varphi = 3.737 \times 0.684 = 2.556 \text{ kVar}$$

Therefore, we can see that the reactive power Q of the asynchronous generator is dependent on the value of the terminal voltage.

6.2.6 Energy Storage System

The charging and discharging control of energy storage is an effective way to improve the power distribution of the system. Grid-scale energy storage could be rated from a few kW to hundreds of MW and many different technologies are used. In this section, the model of a battery is described in detail. The battery model is shown in Figure 6.7.

In the figure, V is the terminal voltage of the ESS, I is the output current, V_0 is the internal voltage source, and R_{eff} is the internal resistance representing the charge-discharge efficiency. The battery has three operating states: charging state, discharging state and idle state [7].

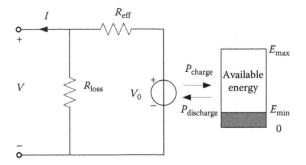

FIGURE 6.7 The model of an ESS.

1. When the battery is charging at a rate of P_{charge}, the energy stored keeps increasing, until it reaches the maximum allowable value (also called the rated capacity) E_{max}. Then the battery will be switched to idle state, and P_{charge} will be zero.

2. Similarly, when the battery is discharging at a rate of $P_{discharge}$, the energy stored keeps decreasing, until it reaches the minimum allowable value E_{min}. Then the battery will be switched to idle state, and $P_{discharge}$ will be zero.

3. When the battery is idle without charging or discharging, there is still a little energy loss, represented by a resistance R_{loss} in the model.

Besides, the charge-discharge power can never be too large to go beyond the limit that the battery can afford, or else it will be switched to idle state.

6.2.7 DSTATCOM

Due to the integration of DGs, the voltage and reactive power problems of distribution systems are becoming increasingly prominent. Reactive power compensation devices, such as distribution static synchronous compensator (DSTATCOM), are used to regulate the voltage at the points of grid connection of DGs [8]. The structure of DSTATCOM is shown in Figure 6.8.

Figure 6.8 shows a schematic diagram of a DSTATCOM system which uses an IGBT-based AC-DC converter. The inverter creates an output AC voltage that is controlled with PWM to produce either leading (capacitive) or lagging (inductive) reactive current into the utility system. The inverter-based DSTATCOM can maintain a constant current during low or

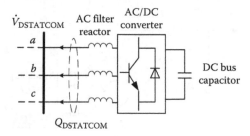

FIGURE 6.8 The schematic of DSTATCOM.

high voltages. Thus, its reactive power is directly proportional to the system voltage.

There are three typical control modes for a DSTATCOM:

6.2.7.1 Q Control Mode

A DSTATCOM in the Q control mode is configured to control the reactive power flow at the point of connection of the DSTATCOM. The bus that the DSTATCOM connects can be regarded as a PQ bus with an active power output $P = 0$ and reactive power output Q equal to a specified value.

6.2.7.2 V Control Mode

A DSTATCOM in the V control mode will maintain the bus voltage at a specified value by controlling the reactive power. Thus, the bus can be regarded as PV bus with active power output $P = 0$ and bus voltage being controlled to a specified value.

6.2.7.3 Droop Control Mode

In this mode, the output voltage of a DSTATCOM is proportional to the output current, namely the output reactive power of the DSTATCOM adopts the droop control. The relationship between the output voltage and output current in droop control is shown in the following equation:

$$I_{STATCOM} = k_{QV} \left(V_{DSTATCOM} - V_{ref} \right) \tag{6.19}$$

where

V_{ref} is the reference voltage of DSTATCOM
$V_{DSTATCOM}$ is the terminal voltage magnitude of DSTATCOM
$I_{DSTATCOM}$ is the output reactive current of DSTATCOM
k_{QV} is the droop slope of DSTATCOM

As shown in Equation 6.19, the relationship between the output reactive power and voltage is shown in Equation 6.20.

$$Q_{STATCOM} = k_{QV} \left(V_{DSTATCOM} - V_{ref} \right) V_{STATCOM} \tag{6.20}$$

It can be seen that the output reactive power is a quadratic function of its terminal voltage, and thus the bus with DSTATCOM can be regarded as a $PQ(V)$ bus with active power output $P = 0$. When the reactive power output of the DSTATCOM reaches its maximum value Q_{max}, the bus with DSTATCOM is converted into a PQ bus with active power output $P = 0$ and reactive power output Q being controlled to Q_{max}.

6.2.8 Dynamic Voltage Regulator

Among many electric power quality issues, the damage caused by voltage fluctuation, such as temporary voltage reduction, increase or flicker, is the most common. The function of a dynamic voltage regulator (DVR) is mainly to compensate voltage sags. It is a powerful means to reduce the influence of voltage fluctuation. Especially with an increasing connection of distribution generations such as wind power, it should take measures to stabilize the voltage at the point of grid connection of DGs and enhance the low voltage ride-through capability of DGs. Under normal conditions, the DVR does not compensate the voltage at the point of grid connection. When faults occur, the DG terminal voltage can be compensated by DVR, enabling the DGs and loads to remain in the grid-connected operation [9]. The single-phase structure of a DVR is shown in Figure 6.9.

The reference voltage for dynamic compensation can be calculated based on the voltage changes caused by daily load changes. The control

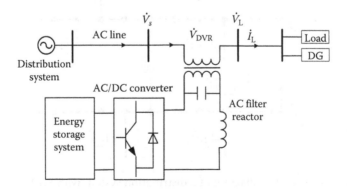

FIGURE 6.9 The single line diagram of a dynamic voltage regulator.

unit monitors the amplitude and phase of the system voltage. When a voltage change is detected, the PWM signals of the inverter and the compensation voltage are given according to the instantaneous value of the system voltage and the reference voltage. The relationship between the compensation voltage and the system node voltage is as follows:

$$\dot{V}_{\mathrm{DVR}} = \dot{V}_{\mathrm{S}} - \dot{V}_{\mathrm{L}} \tag{6.21}$$

In steady-state analysis, a DVR can be regarded as an ideal voltage source connected to the distribution system.

6.2.9 SOP

A soft open point (SOP) is a multi-purpose power electronic device that provides active and reactive power management functions in distribution systems. The distribution systems are typically classified as radial or meshed systems, each with their own inherent advantages and disadvantages. By placing SOPs into a distribution system, a hybrid system type is formed with the advantages of both radial and meshed systems. There are also many other key features and properties of a distribution system with SOPs, such as reduced total network losses and improved reliability of existing distribution networks [10,11]. The single line diagram of a simple distribution system with two feeders and one SOP connecting them is shown in Figure 6.10.

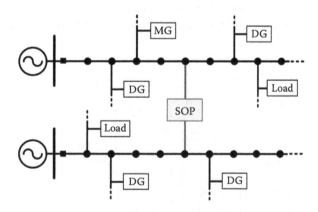

FIGURE 6.10 Single line diagram of a distribution system with SOP.

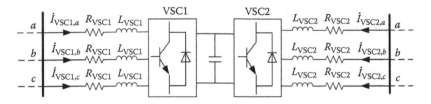

FIGURE 6.11 Topology structure of a B2B VSC–based SOP.

SOPs enable the interconnection of multiple feeders, coordinated control among SOPs and substations for load balancing, and fault isolation of SOP-connected feeders. A back-to-back voltage source converter (B2B VSC)-based SOP is shown in Figure 6.11. Distribution-level unified power flow controllers (D-UPFC) and static synchronous series compensators (SSSC) can also be used as SOPs.

In a B2B VSC–based SOP, the operating point of each converter can be adjusted by its own control system. Under the normal operating conditions of a distribution system, one converter needs to keep the DC voltage stable and the other to control the active power. Control modes 1–4 are listed in Table 6.2, with which the voltage at the converter terminals is regulated to a certain level or reactive power set points are assigned as required by the converter control system. The corresponding node type is to be treated as a *PQ* or a *PV* node.

TABLE 6.2 SOP Control Mode

Control Mode	VSC1 Control	Node Type	VSC2 Control	Node Type
1	*PQ* control	*PQ*	$V_{DC}Q$ control	*PQ*
2	*PQ* control	*PQ*	$V_{DC}V_{AC}$ control	*PV*
3	PV_{AC} control	*PV*	$V_{DC}Q$ control	*PQ*
4	PV_{AC} control	*PV*	$V_{DC}V_{AC}$ control	*PV*
5	$V_{AC}\theta$ control	*Vθ*	$V_{DC}Q$ control	*PQ*
6	$V_{AC}\theta$ control	*Vθ*	$V_{DC}V_{AC}$ control	*PV*

In an SOP, there are conduction and switching losses. The converter losses can be simply described by Equation 6.22.

$$P_{VSC1} + \left(I^2_{VSC1,a} + I^2_{VSC1,b} + I^2_{VSC1,c}\right)R_{VSC1} + \left(I^2_{VSC2,a} + I^2_{VSC2,b} + I^2_{VSC2,c}\right)R_{VSC2} + P_{VSC2} = 0$$

$$(6.22)$$

where

P_{VSC1} and P_{VSC2} are the injected active power of two converters into the AC distribution system

$I_{VSC1,a}$, $I_{VSC1,b}$, $I_{VSC1,c}$ and $I_{VSC2,a}$, $I_{VSC2,b}$, $I_{VSC2,c}$ are the injected three-phase currents of two converters

R_{VSC1} and R_{VSC2} are the equivalent resistance of two converters, including resistance of switches, filters and other components of the converters

In power flow analysis, the injected currents of two converters are calculated by using the injected active power. Equation 6.22 is used to obtain the injected active power of one converter when the injected active power of the other converter is known based on the control mode.

When a fault occurs in the distribution system, the protection device will isolate the fault, and some loads may be interrupted, as shown in Figure 6.12. Under this condition, control mode 5 or 6 is activated to maintain the power supply of these loads with $V_{AC}\theta$ control of VSC1.

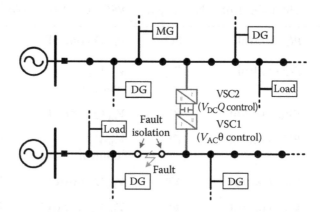

FIGURE 6.12 SOP control mode under fault condition.

6.3 Controllers for Network Components

6.3.1 Voltage Regulator

The function of a voltage regulator is to control the tap position of a transformer, regulate the system voltage and increase the operation efficiency. The main data acquired are sampled voltage and current values of the potential transformer (PT) and the current transformer (CT), as shown in Figure 6.13.

　　The choice of average voltage, maximum voltage, minimum voltage and single-phase voltage (phase *a*, *b* or *c*) depends on different requirements. They are defined as

1. Average voltage

$$V_{mea} = \frac{|V_a + V_b + V_c|}{k} \tag{6.23}$$

2. Maximum voltage

$$V_{mea} = \frac{\max(V_a, V_b, V_c)}{k} \tag{6.24}$$

3. Minimum voltage

$$V_{mea} = \frac{\min(V_a, V_b, V_c)}{k} \tag{6.25}$$

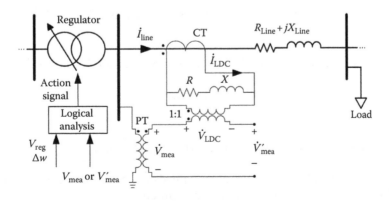

FIGURE 6.13 The control logic of a voltage regulator.

4. Single-phase voltage (e.g. phase a)

$$V_{\text{mea}} = \frac{V_a}{k} \tag{6.26}$$

where

V_a, V_b and V_c are the three-phase secondary voltages of the transformer to which the PT connects

k is the ratio of PT

There are two ways to adjust the voltage: direct control and indirect control. In direct control, the regulator controls the voltage of one specified bus directly. First, the allowed range for the bus voltage is set as $V_{\text{reg}} - (1/2) \times \Delta w \sim V_{\text{reg}} + (1/2) \times \Delta w$, where V_{reg} is the voltage reference value and Δw is the allowed variation. When the sampled voltage V_{mea} goes out of range, the tap position is adjusted. The tap step is calculated as follows:

1. Calculate the difference V_{boost} between the sampled voltage and the reference value:

$$V_{\text{boost}} = \begin{cases} V_{\text{reg}} - \dfrac{\Delta w}{2} - V_{\text{mea}}, & V_{\text{mea}} < V_{\text{reg}} - \dfrac{\Delta w}{2} \\ V_{\text{mea}} - \left(V_{\text{reg}} + \dfrac{\Delta w}{2} \right), & V_{\text{mea}} > V_{\text{reg}} + \dfrac{\Delta w}{2} \end{cases} \tag{6.27}$$

2. Obtain the per unit value of V_{boost}:

$$V_{\text{boost}}^* = \frac{V_{\text{boost}} \times k}{V} \tag{6.28}$$

where V is the rated value of the secondary winding voltage of the transformer.

3. Calculate the number of times the tap position needs to be adjusted:

$$n = \left\lfloor \frac{V_{\text{boost}}^*}{h} \right\rfloor \tag{6.29}$$

where h is the change in voltage with one tap step.

4. Since the number n is an integer, n needs to be rounded first, and then the variation of the tap ratio t_c is computed.

$$t_c = n \times h \tag{6.30}$$

According to the calculated variation t_c, determine whether the current tap position meets the requirements of the adjustment. The adjusted tap position shall not exceed the range of the largest tap and the lowest tap. If the condition is met, the regulator sends out the control signal to change the tap position.

For the direct control, the sampled voltage can be local nodal voltage, but it can also be a remote nodal voltage, which is transferred using the communication system. For indirect control, the regulator does not need to sample the remote nodal voltage, but the local voltage and the position of the tap is determined by a control circuit (line drop compensator). Line drop compensator uses resistance R and reactance X to simulate the parameters of downstream lines. Equation 6.31 gives the voltage $\dot{V}'_{mea}$ corrected by the compensation circuit. Equation 6.27 is changed to Equation 6.32, replacing $\dot{V}'_{mea}$.

$$\dot{V}'_{mea} = \dot{V}_{mea} - (R + jX)\dot{I}_{LDC} \tag{6.31}$$

$$V_{boost} = \begin{cases} V_{reg} - \dfrac{\Delta w}{2} - \dot{V}'_{mea}, & \dot{V}'_{mea} < V_{reg} - \dfrac{\Delta w}{2} \\[2ex] \dot{V}'_{mea} - \left(V_{reg} + \dfrac{\Delta w}{2}\right), & \dot{V}'_{mea} > V_{reg} + \dfrac{\Delta w}{2} \end{cases} \tag{6.32}$$

Example 6.6

A transformer rated at 20 MVA, 110/10.5 kV is controlled by the voltage regulator in Figure 6.13. The tap is $1 \pm 8 \times 0.625\%$ and set on the 0 position. The equivalent line impedance between the regulator and the load centre node is $R_{Line} + jX_{Line} = 0.3 + j0.9\ \Omega$. The R and X settings in the compensator circuit is $R + jX = 1.438 + j4.313\ \Omega$. The PT ratio connected to the compensator circuit is 48. The CT ratio is 230 and the rated current of the primary side is 1150 A. The voltage level of the regulator has been set at 120 V with a bandwidth of 2 V. Assuming that the load is rated at 10 kV, 9 MW with a power factor of 0.9, calculate what tap the regulator should move to?

Answer

The current flowing through the line and the line drop compensator are

$$\dot{I}_{line} = \frac{9000}{\sqrt{3} \times 10} \angle \left(-\cos^{-1} 0.9 \right) = 519.615 \angle - 25.842\,\text{A}$$

$$\dot{I}_{LDC} = \frac{519.615 \angle - 25.842}{230} = 2.259 \angle - 25.842\,\text{A}$$

The sampled voltage corrected by the compensation circuit based on Equation 6.31 is

$$\dot{V}'_{mea} = \dot{V}_{mea} - \left(R + jX \right) \dot{I}_{LDC}$$
$$= \frac{10,000 / \sqrt{3}}{48} \angle 0 - \left(1.438 + j4.313 \right) \times 2.259 \angle - 25.842$$
$$= 113.349 \angle - 3.719\,\text{V}$$

According to Equation 6.32, the difference between the sample voltage and the reference value is calculated as

$$V_{boost} = 119 - 113.349 = 5.651\,\text{V}$$

Then V^{*}_{boost} is given by

$$V^{*}_{boost} = \frac{V_{boost} \cdot k}{V} = \frac{5.651 \times 48}{10,000 / \sqrt{3}} = 0.04698$$

The number of the tap position needs to be adjusted is

$$n = \left\lfloor \frac{V^{*}_{boost}}{h} \right\rfloor = \left\lfloor \frac{0.04698}{0.00625} \right\rfloor = 7.5168 = 7$$

The number n is rounded to 7. This means the tap needs to be adjusted to the +7 position from the current 0 position.

6.3.2 Capacitor Controller

The function of a capacitor controller is to control the switching of the capacitor. The capacitor controller first detects the power flow data such as current, voltage and power. Then these data are analysed and judged whether the capacitor needs to be switched on or off. Finally, the controller sends out the action signal. A model of the capacitor controller is shown in Figure 6.14.

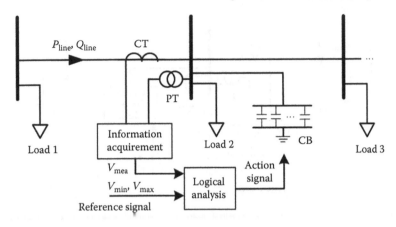

FIGURE 6.14 The control logic of a capacitor controller.

The voltage of one bus is taken as an example of the sample variable to illustrate the control process. Normally, this bus is the capacitor connection point. The capacitor controller determines whether the number of capacitors needs to be increased or decreased by comparing the sample voltage V_{mea} and the reference value. It is assumed that the allowed range of the sampled voltage is $V_{min} \sim V_{max}$.

1. When V_{mea} is less than V_{min}, the controller will send out a 'switching-on' signal to the capacitor, so that the voltage can be increased.

2. When V_{mea} is larger than V_{max}, the controller will send out a 'switching-off' signal to the capacitor, so that the voltage can be decreased.

Similarly, the capacitor controller can control the switching of capacitors based on other power flow data such as current, reactive power and power factor. The capacitors will be switched on or off until the sampled variable meets the operation requirement or the capacitor group has been all switched on or off.

6.3.3 Energy Storage Controller

Using a controller, energy storage can be controlled under a schedule or a mode, as shown in Figure 6.15.

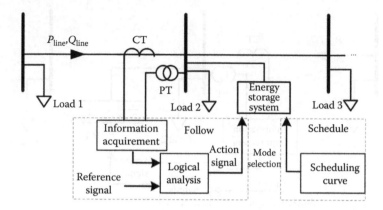

FIGURE 6.15 Two control modes of an energy storage controller.

1. *Schedule mode*: Energy storage controller receives the scheduling signal from the distribution network operation centre, which has been optimized by the distribution management system using the load and DG forecasting information.

2. *Following mode*: The logical controller analyses the information acquired by the local CT and PT and sends action signal to the ESS. Then the energy storage charging and discharging power is changed accordingly.

Under the following mode, the local CT and PT check the voltage of one bus or the active power flow on one line nearby as input to the ESS controller. Taking the line active power for example setting the control band between P_{min} and P_{max}, means that once the active power breaches this band, the controller will adjust the active power of the ESS to maintain the line active power within this band.

6.4 Power Flow Analysis

The main purpose of power flow analysis is to obtain the current operation status of a distribution network. Essentially the problem is to solve a group of non-linear equations. The nodal voltage equation is expressed as follows.

$$\dot{I} = Y\dot{V} \tag{6.33}$$

where
$\dot{I}$ represents the injection current vector
$\dot{V}$ is the voltage vector
Y is the nodal admittance matrix of the entire system, which is formed by the admittance matrix of lines, transformers and other components

The current injected into each node is related to the power equipment connected. For the wye-connected and constant real and reactive power (constant *PQ*) load, the injected current is calculated as follows, where *i* represents the node number and *p* represents the phase *a*, *b* or *c*.

$$\dot{I}_{ip} = -\left(\frac{P_{ip} + jQ_{ip}}{\dot{V}_{ip}} \right)^* \qquad (6.34)$$

Substitute Equation 6.34 into Equation 6.33 to obtain a set of non-linear equations. The voltage of each node can be obtained with the equations solved, and thus we can know about the power flow distribution of the entire system.

The commonly used methods to solve power flow equations are the Zbus Gauss method, the Newton method, the improved fast decoupled method, the backward/forward sweep method and the loop-impedance method [12,13]. These algorithms are distinct in their convergence, efficiency, applicability and programmability. Here, the Zbus Gauss method is mainly introduced.

Separating the slack bus from other buses, the power flow equation becomes

$$\begin{bmatrix} \dot{I}_1 \\ \dot{I}_2 \end{bmatrix} = \begin{bmatrix} Y_{11} & Y_{12} \\ Y_{21} & Y_{22} \end{bmatrix} \begin{bmatrix} \dot{V}_1 \\ \dot{V}_2 \end{bmatrix} \qquad (6.35)$$

where
  $\dot{I}_1$ and $\dot{V}_1$ are the current and voltage vectors of the slack bus
  $\dot{I}_2$ and $\dot{V}_2$ are that of the others

For a distribution system, the voltage vector of the slack bus $\dot{V}_1$ is generally given. If $\dot{I}_2$ is known as constant, then the voltages of all nodes except the slack bus can be solved as follows:

$$\dot{V}_2 = Y_{22}^{-1} \left[\dot{I}_2 - Y_{21}\dot{V}_1 \right] \qquad (6.36)$$

If the injected current changes with the node voltage (such as the currents of constant *PQ* loads), one of the results of the power flow analysis is to replace the actual voltages with the assumed voltages. The current vector will be a function of the node voltage vector. In the Gauss iterative algorithm, the voltage always uses the result $\dot{V}_2$ of the previous iteration in every iteration, as shown in Equation 6.37:

$$\dot{V}_2^{(k)} = Y_{22}^{-1} \left[\dot{I}_2 \left(\dot{V}_2^{(k-1)} \right) - Y_{21}\dot{V}_1 \right] \qquad (6.37)$$

The algorithm ends when the voltage change between two iterations meets the precision requirement.

This iterative process shows that the algorithm is constantly multiplying an impedance matrix Y_{22}^{-1}, so this algorithm is called the Zbus Gauss method. Y_{22}^{-1} does not need to be explicit by the implementation of Y_{22}'s factor table, so the Zbus Gauss method is an implicit algorithm. The implementation of the Zbus Gauss method consists of two stages: calculating $\dot{I}_2\left(\dot{V}_2^{(k-1)}\right)$ and the backward/forward sweep of Equation 6.37 using Y_{22}'s factor table. The calculation procedure is as follows:

1. Input base data and initialize the voltage of each node.

2. Add shunt capacitors and constant impedance loads so that the nodal admittance matrix Y is formed and stored.

3. Obtain Y_{22} by separating the slack bus from other nodes.

4. Carry out factorization of Y_{22}.

5. Calculate the node current vector $\dot{I}_2$ except the slack bus using the nodal voltages of the previous iteration.

6. Use the Gauss iterative method to get the value of $\dot{V}_2$ by solving Equation 6.37.

7. Calculate the voltage difference of each node, and determine whether the algorithm converges by comparing the difference to the precision. If it converges, the iteration ends; if not, repeat step 5.

The Zbus Gauss method is a simple power flow algorithm which needs less computer memory and is suitable for most distribution systems. As a first-order method, the Zbus Gauss method can ensure a good convergence property and have a fast rate of convergence.

For DGs modelled as *PV, PI* and *PQ(V)* buses, they need to be converted into *PQ* buses in the Zbus Gauss method.

First, the real power P and voltage magnitude V of the *PV* buses are constant. It can be treated as *PQ* buses in power flow program where its reactive power is initially assumed. If the calculated voltage magnitude is different as predefined, the reactive power needs to be adjusted by Equation 6.38:

$$Q_{k+1} = Q_k + X^{-1}V_k\Delta V_k \qquad (6.38)$$

where

X is the imaginary component of the system's equivalent impedance seen from the DG node

V_k is the voltage magnitude of the DG node in the kth iteration

ΔV_k is the difference between the voltage magnitude in the current iteration and the reference value

Second, the reactive power of a PI bus is calculated by the nodal voltage of the previous iteration, the current magnitude and its real power, as shown in Equation 6.39.

$$Q_{k+1} = \sqrt{|I|^2 V_k^2 - P^2} \qquad (6.39)$$

where

Q_{k+1} is the reactive power of the DG

V_k is the voltage magnitude of the PI bus in the kth iteration

I is the constant current magnitude of the DG

P is the constant real power

Finally, the treatment to the $PQ(V)$ bus is similar to the PI bus. Take the asynchronous generator as an example. Its reactive power can be calculated using the voltage of the previous iteration as well as Equation 6.39. Then the bus that the asynchronous generator connects to is treated as a PQ bus in the current iteration.

Example 6.7

Figure 6.16 shows a simple 4-node distribution network rated at 110 kV. The corresponding parameters are as follows:

a. Node B0 is connected to an infinite source, so it is considered as the slack bus. The three-phase voltages of node B0 to node B3 are assumed to be $\dot{V}_{0,p} \sim \dot{V}_{3,p}$ ($p = a, b, c$), respectively.
b. The lengths of Line 2 and Line 3 are 2 and 2.5 km, respectively. The parameters of lines are the same as those in Example 6.1. Their admittance matrixes are Y_{12} and Y_{23}, which are equal to the inverse of their series impedance matrixes without considering their line-to-ground capacitance.

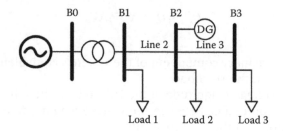

FIGURE 6.16 The simple 4-node example.

    c.   The transformer is delta-grounded wye step-down con-
nected as shown in Example 6.2 and its admittance matrix is
Y_{01} (its four block matrixes are Y_{01}^{11}, Y_{01}^{12}, Y_{01}^{21} and Y_{01}^{22}).

  d. Load 1, Load 2 and Load 3 are all wye connected, three-phase
balanced and constant PQ loads. The real power $P_{1,p} \sim P_{3,p}$ are
0.5, 1 and 2 MW, respectively, and the power factors are all
0.9. The reactive power is $Q_{1,p} \sim Q_{3,p}$, respectively.

  e. A PV station is wye connected to node B2 and constant PQ
with the real power of $P_{DG,p}$ and reactive power of $Q_{DG,p}$, which
contains 8 PV systems. The rated capacity of each PV inverter
is 0.5 MVA and other parameters are shown in Example 6.4.

    Derive the power flow equation and calculate the voltage
of each node.

Answer

Kirchhoff's current law (KCL) and voltage law (KVL) are applied
to obtain the power flow equation:

$$
\begin{bmatrix}
[Y_{01}^{11}]_{3\times3} & [Y_{01}^{12}]_{3\times3} & 0 & 0 \\
[Y_{01}^{21}]_{3\times3} & [Y_{01}^{22}]_{3\times3}+[Y_{12}]_{3\times3} & -[Y_{12}]_{3\times3} & 0 \\
0 & -[Y_{12}]_{3\times3} & [Y_{12}]_{3\times3}+[Y_{23}]_{3\times3} & -[Y_{23}]_{3\times3} \\
0 & 0 & -[Y_{23}]_{3\times3} & [Y_{23}]_{3\times3}
\end{bmatrix}
\dot{V}=\dot{I}
$$

where

$$
\dot{V}=\begin{bmatrix} \dot{V}_{0a} & \dot{V}_{0b} & \dot{V}_{0c} & \dot{V}_{1a} & \dot{V}_{1b} & \dot{V}_{1c} & \dot{V}_{2a} & \dot{V}_{2b} & \dot{V}_{2c} & \dot{V}_{3a} & \dot{V}_{3b} & \dot{V}_{3c} \end{bmatrix}^{T}
$$

$$
\dot{I}=\begin{bmatrix} \dot{I}_{0a} & \dot{I}_{0b} & \dot{I}_{0c} & \dot{I}_{1a} & \dot{I}_{1b} & \dot{I}_{1c} & \dot{I}_{2a} & \dot{I}_{2b} & \dot{I}_{2c} & \dot{I}_{3a} & \dot{I}_{3b} & \dot{I}_{3c} \end{bmatrix}^{T}
$$

The admittance matrix of the transformer and the series imped-ance matrix of Line 2 have been calculated in the previous exam-ples. Since the length of Line 3 is 2.5 km, its series impedance matrix needs to be recalculated:

$$Z_{\text{Line3}} = \begin{bmatrix} 1.375 + j1.817 & 0.550 + j0.727 & 0.550 + j0.727 \\ 0.550 + j0.727 & 1.375 + j1.817 & 0.550 + j0.727 \\ 0.550 + j0.727 & 0.550 + j0.727 & 1.375 + j1.817 \end{bmatrix}$$

The admittance matrix is obtained through matrix inversion. Substitute the admittance matrixes of the lines and transformer into power flow equation and the admittance matrix Y_{system} of the entire system can be obtained as follows.

Then the system power flow equation is established and solved using the Zbus Gauss method. The load currents are given by

$$\dot{I}_{ip} = -\left(\frac{P_{ip} + jQ_{ip}}{\dot{V}_{ip}} \right)^*, \quad i = 1, 2, 3$$

The DG current is given by

$$\dot{I}_{2,p} = \left(\frac{P_{DG,p} + jQ_{DG,p}}{\dot{V}_{2,p}} \right)^*$$

where
i is the node number and p represents the phase a, b or c
$\dot{V}_{2,p}$ is the given voltage of the slack bus

This formula collectively contains nine equations and nine unknowns to be solved. Then the calculation procedure detailed before is executed:

1. Initialize the voltage of each node, as shown in Table 6.3.
2. Obtain the value of $\dot{V}_2$ by solving Equation 6.37; the first itera-tion results of nodal voltages are shown in Table 6.3;
3. Calculate the voltage difference of each node. The maximum voltage difference is 0.5401766 at node B2, exceeding the solution toleration 1.0E−7.
4. Repeat the said Gauss iterative method until the solution converges.

    The detailed iteration results and final nodal voltages are shown in Table 6.3.

TABLE 6.3 Nodal Voltage of the 4-node Distribution Network

Node Number	Phase	Voltage under Each Iteration (kV)					
		0 (Initial)	1	2	...	...	7 (Final)
B0	A	63.509∠0.0	63.509∠0.0	63.509∠0.0	...	...	63.509∠0.0
	B	63.509∠−120.0	63.509∠−120.0	63.509∠−120.0	...	...	63.509∠−120.0
	C	63.509∠120.0	63.509∠120.0	63.509∠120.0	...	...	63.509∠120.0
B1	A	6.062∠0.0	6.024∠30.4	6.006∠30.1	...	...	6.004∠30.2
	B	6.062∠−120.0	6.024∠−89.6	6.006∠−89.9	...	...	6.004∠−89.8
	C	6.062∠120.0	6.024∠150.4	6.006∠150.1	...	...	6.004∠150.2
B2	A	6.062∠0.0	6.045∠31.4	5.967∠31.1	...	...	5.964∠31.1
	B	6.062∠−120.0	6.045∠−88.6	5.967∠−88.9	...	...	5.964∠−88.9
	C	6.062∠120.0	6.045∠151.4	5.967∠151.1	...	...	5.964∠151.1
B3	A	6.062∠0.0	5.879∠31.5	5.815∠30.3	...	...	5.809∠30.4
	B	6.062∠−120.0	5.879∠−88.5	5.815∠−89.7	...	...	5.809∠−89.6
	C	6.062∠120.0	5.879∠151.5	5.815∠150.3	...	...	5.809∠150.4

6.5 Sequential Power Flow

System loads and generation change over time. With more and more DGs integrated into the grid, the randomness and volatility of the system will be more obvious. For wind turbines and PV systems, the wind speed and irradiance may fluctuate greatly in seconds or minutes. At the same time, there are many control components in the grid to ensure the system is operating within the limits by adjusting the power flow. This is where sequential power flow comes into play. It enables sequentially simulation of the operation states or problems of a distribution power system. Then different control strategies can be used to improve the operational condition [14,15].

Sequential power flow is divided into yearly power flow analysis, daily power flow analysis and short-term power flow analysis. Yearly and daily power flow are widely used, but they cannot capture the variations in DG output. The short-term power flow is mainly dedicated to analyzing the changes of loads or DGs in seconds or minutes with a shorter time step.

The framework of sequential power flow is shown in Figure 6.17. Sequential power flow is based on the basic power flow calculation. Not only are the basic information such as grid topology and parameters of components needed, but the necessary operation data, such as the operation curves of loads and the variation of irradiance, temperature, and wind speed are also required. Through sequential power flow, we can

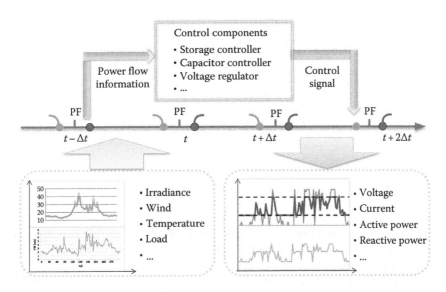

FIGURE 6.17 The framework of sequential power flow in a smart distribution network.

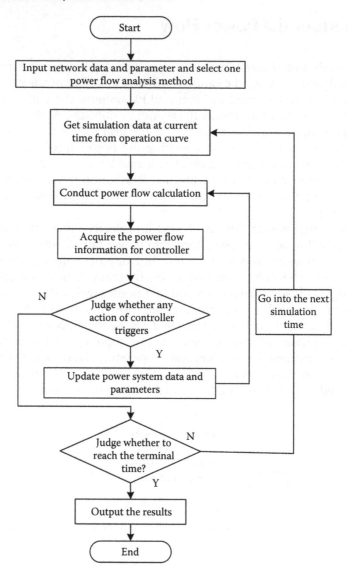

FIGURE 6.18 The flow chart of sequential power flow in a smart distribution network.

obtain the operating conditions of a distribution system for the daily or yearly period and analyse the effect of variation in the DG output on the distribution system.

Besides, sequential power flow also pays particular attention to different controllers associated with energy storage, switches, capacitors and voltage regulators. These control components first acquire power flow

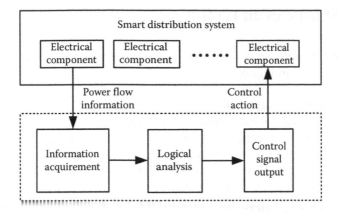

FIGURE 6.19 The structure of typical control components.

data of the current distribution system, then analyse the control strategy and send out control action signals and finally make adjustments to the electrical components in the distribution system and improve the operational condition. The overall process of sequential power flow is given in Figure 6.18.

Using the models of control components, we can conduct a closed-loop simulation of the distribution system, as shown in Figure 6.19. The typical control components mainly contain three modules: information acquirement, logical analysis and control signal output. The current power flow information can be obtained by data collection, then the control signal is sent out based on the corresponding control strategy. The purpose of control components is to adjust the operation mode of the primary electrical components of the distribution system. A detailed description of the most common control components is given in Figure 6.19.

6.6 Short-Circuit Fault Analysis

Short-circuit faults occur in the power system due to many reasons, including tree branches falling on power lines, natural disasters and animals coming into contact with power lines. A short-circuit fault may lead to very serious consequences, such as current surge and damage to power system equipment and even to humans. Therefore, short-circuit fault analysis is important for the planning, design and operation management of a smart distribution system.

6.6.1 Short-Circuit Fault Types

Short-circuit faults in the distribution system can be divided into four types, as shown in Figure 6.20.

1. Three-phase to earth fault

    A three-phase to earth short-circuit fault in a three-phase system is a balanced or symmetrical fault. When it occurs, as a bolted fault, every phase voltage at the fault point becomes equal or close to zero. With a non-bolted grounding fault, the voltage is a certain value greater than zero but less than the rated phase voltage.

2. One-phase to earth fault

    When a one-phase to earth short-circuit fault occurs in a power distribution system, for a bolted fault, the voltage at the fault point is equal or close to zero, and the other two-phase voltages are approximately equal to the rated voltage. For a non-bolted fault, the voltage is a certain value greater than zero but less than the rated phase voltage, and the other two-phase voltages are close to the rated phase voltage.

3. Phase-to-phase fault

    When this type of fault occurs, the short-circuit currents of two faulty phases are opposite to each other, and the current on the healthy phase is very small compared to the faulty phases. With a bolted fault, the voltages of the two faulted phases are equal to each other.

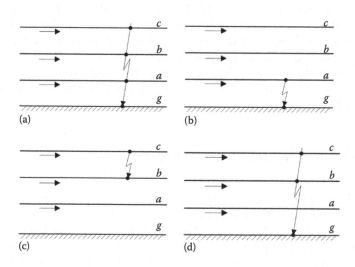

FIGURE 6.20 Four types of short-circuit faults: (a) three-phase to earth fault, (b) one-phase to earth fault, (c) phase-to-phase fault and (d) two-phase to earth fault.

4. Two-phase to earth fault

When this type of fault occurs, for a bolted fault, the voltages at the fault point are equal or close to zero, and the other phase voltages are approximately equal to the rated voltage. For a non-bolted fault, each faulted phase voltage is a certain value greater than zero but less than the rated phase voltage, and the other two-phase voltages are close to the rated phase voltage.

6.6.2 Methodology of Fault Analysis

Several methods are employed in the fault analysis of distribution systems, including the equivalent voltage source method, the equivalent element method and the transient simulation method. These methods provide different applicability in various situations of fault analysis. In this section, the equivalent element approach is taken as an example to illustrate the basic ideas.

The equivalent element method is developed as an extension of power flow calculation. It is capable of dealing with all types of faults with high accuracy and low programming cost. The following four steps are used in this method:

1. An equivalent circuit is adopted to simulate the fault that occurs.

2. The corresponding elements in the admittance matrix for power flow calculation are altered according to the equivalent fault circuit.

3. The steady-state fault current is derived by solving the new power flow equation.

4. The instantaneous value of the fault current is calculated based on its steady-state values.

For any unbalanced fault, the fault circuit can always be separated from the main network at the faulty point and be represented by a compensation circuit, as shown in Figure 6.21.

According to Figure 6.21, the fault impedance matrix is written as

$$Z_f = \begin{bmatrix} z_{fg} + z_{fa} & z_{fg} & z_{fg} \\ z_{fg} & z_{fg} + z_{fb} & z_{fg} \\ z_{fg} & z_{fg} & z_{fg} + z_{fc} \end{bmatrix} \tag{6.40}$$

where

z_{fa}, z_{fb} and z_{fc} are the three-phase virtual impedance, indicating the fault impedance among phases

z_{fg} is the virtual ground impedance, indicating the fault resistance between phase and ground

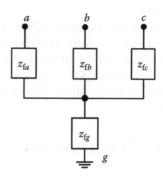

FIGURE 6.21 Separation of a fault circuit.

All fault types, including bolted and non-bolted short-circuit faults, can be simulated by setting proper values for z_{fa}, z_{fb}, z_{fc} and z_{fg}. Several typical values for bolted faults are given in Table 6.4. Besides, the values of z_{fa}, z_{fb}, z_{fc} and z_{fg} can be set to any value between 0 and 10^6, in order to represent non-bolted faults.

TABLE 6.4 Impedance for Typical Bolted Faults

Fault Type	Impedance Value
One-phase to earth fault on phase a	$z_{fa} = z_{fg} = 0$, $z_{fb} = z_{fc} = 10^6$
Two-phase fault on phases a and b	$z_{fa} = z_{fb} = 0$, $z_{fc} = z_{fg} = 10^6$
Two-phase to earth fault on phases a and b	$z_{fa} = z_{fb} = z_{fg} = 0$, $z_{fc} = 10^6$
Three-phase fault	$z_{fa} = z_{fb} = z_{fc} = 0$, $z_{fg} = 10^6$
Three-phase to earth fault	$z_{fa} = z_{fb} = z_{fc} = z_{fg} = 0$

Example 6.8

A three-phase non-bolted fault occurs at point K in Line 3, as shown in Figure 6.22. The grounding impedance is 5 Ω. Write the fault impedance matrix for fault analysis.

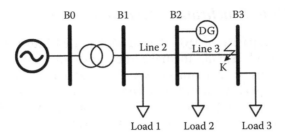

FIGURE 6.22 Three-phase, non-bolted grounding fault.

Answer

The fault is a non-bolted grounding fault, so $z_{fg} = 5\ \Omega$, $z_{fa} = z_{fb} = z_{fc} = 0\ \Omega$. According to Equation 6.40, the impedance matrix can be written as

$$
Z_f = \begin{bmatrix} z_{fg} + z_{fa} & z_{fg} & z_{fg} \\ z_{fg} & z_{fg} + z_{fb} & z_{fg} \\ z_{fg} & z_{fg} & z_{fg} + z_{fc} \end{bmatrix} = \begin{bmatrix} 5 & 5 & 5 \\ 5 & 5 & 5 \\ 5 & 5 & 5 \end{bmatrix} \Omega
$$

The fault admittance matrix is derived by inverting the impedance matrix. The fault admittance matrix and the system admittance matrix are combined together according to the node number of the faulty point. Then fault calculation is performed with the integrated admittance matrix, based on the basic theory of power flow calculation.

6.6.3 Fault Analysis Considering Distributed Generators

The terminal voltage of a DG decreases when a short-circuit fault occurs in the system. The low voltage ride-through capability allows the DG to support the operation of the distribution network under fault conditions. DG units without the capability of low voltage ride-through will be shut down. Hence, only the DGs that keep operating are considered in the fault analysis.

Various DGs are integrated into distribution systems. They are divided into three types according to their integration interfaces, including the synchronous generator type, the asynchronous generator type and the power electronic type. Different models should be designed in short-circuit calculation for DGs of different types [16].

6.6.3.1 DGs of Synchronous Generator Type

Generally, the capacity of DGs that are integrated into the distribution system is relatively small. To make the short-circuit calculation more practical, the transient dynamics of synchronous generators must be considered. For simplicity, the sub-transient reactance on d- and q-axis are assumed to be equal. Taking the winding damping into consideration, the voltage equation of synchronous generator is written as

$$\dot{E}'' = \dot{V} + j\dot{I}X''_d + \dot{I}R_a \tag{6.41}$$

where

$\dot{E}''$ represents the sub-transient electromotive force

R_a represents the stator resistance

X''_d represents the sub-transient reactance on the d-axis

$\dot{V}$ and $\dot{I}$ denote the terminal voltage and the output current of the generator

The sub-transient equivalent circuit of a synchronous generator is shown in Figure 6.23.

The value of sub-transient electromotive force $\dot{E}''$ is constant at the moment when fault occurs. Therefore, the value of $\dot{E}''$ after fault can be derived from the power flow result before fault, as shown in the following equation:

$$\dot{E}'' = \dot{E}''_{0-} = \dot{V}_{0-} + j\dot{I}_{0-}X''_d + \dot{I}_{0-}R_a \tag{6.42}$$

where

$\dot{E}''_{0-}$ is the sub-transient electromotive force before fault

$\dot{V}_{0-}$ is the terminal voltage of generator before fault

$\dot{I}_{0-}$ is the output current of generator before fault

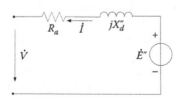

FIGURE 6.23 Sub-transient equivalent circuit of a synchronous generator.

Performing phase-sequence transformation for Equation 6.43, the equivalent inner electromotive force of the synchronous generator is then obtained:

$$\begin{cases} \dot{E}_a = \dot{E}'' \\ \dot{E}_b = e^{j240°} \dot{E}'' \\ \dot{E}_c = e^{j120°} \dot{E}'' \end{cases} \tag{6.43}$$

Furthermore, considering the terminal voltage of the synchronous generator, the output current at the moment when fault occurs can be written as

$$\begin{cases} \dot{I}_a = \dfrac{\dot{E}_a - \dot{V}_a}{R_a + jX_d''} \\[2mm] \dot{I}_b = \dfrac{\dot{E}_b - \dot{V}_b}{R_a + jX_d''} \\[2mm] \dot{I}_c = \dfrac{\dot{E}_c - \dot{V}_c}{R_a + jX_d''} \end{cases} \tag{6.44}$$

6.6.3.2 DGs of Asynchronous Generator Type

An asynchronous generator can be considered as a special case of a synchronous generator. That is, an asynchronous generator is the same as a synchronous generator that rotates at an asynchronous speed with no excitation voltage and equal d- and q-axis parameters. Similar to synchronous generators, the flux linkage of stator and rotor windings in an asynchronous generator must be continuous when an external fault occurs. Thus, the DC components appear in the induced currents in both stator and rotor windings after a fault. Therefore, a sub-transient electromotive force $\dot{E}''$, which is proportionate to the rotor flux, is defined as the equivalent electromotive force in the stator transients of asynchronous generators. With the equivalent sub-transient reactance X'' and the stator resistance R_a, the equivalent circuit of asynchronous generators is derived in a similar way to that for the synchronous generators. The sub-transient reactance X'' can be expressed as

$$X'' = X_{a\sigma} + \frac{X_{r\sigma} X_{ad}}{X_{r\sigma} + X_{ad}} \tag{6.45}$$

where
$X_{a\sigma}$ denotes the stator leakage reactance
$X_{r\sigma}$ denotes the rotor leakage reactance
X_{ad} denotes the reactance of d-axis armature reaction, which is equal to the excitation reactance physically

The sub-transient electromotive force $\dot{E}''$ is also constant at the moment when the fault occurs, which can be calculated from the power flow results before the fault. The short-circuit calculation with DGs based on asynchronous generators is the same as that with synchronous-generator-based DGs.

6.6.3.3 DGs of Power Electronic Type

Some DGs, such as PV systems, fuel cell systems and batteries, are integrated into the distribution network via power electronic inverters. These power inverters can be divided into two groups, including the voltage source inverter and the current source inverter.

For current source inverters, the output current during a transient process is assumed to be constant [17]. Hence, the current source inverters are equivalent to current sources in fault calculations, as shown in Equation 6.46.

$$\dot{I}_{\text{inv}} = \dot{I}_{\text{inv},0^-} \tag{6.46}$$

where
    $\dot{I}_{\text{inv}}$ denotes the inverter output current
    $\dot{I}_{\text{inv},0^-}$ denotes the output current before the fault occurs

The configuration of voltage source inverters is shown in Figure 6.24. The terminal voltage $\dot{E}$ maintains the same value before and after fault. Therefore, the terminal voltage can be derived from the power flow result before fault occurs as long as the active power output P and reactive power output Q are known quantities. Then, phase-sequence transformation is adopted to calculate the three-phase terminal voltage of the inverter. On this basis, the injection current of the inverter under fault

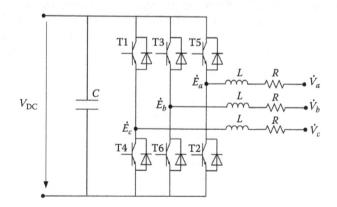

FIGURE 6.24 Main circuit configuration of voltage source inverter.

conditions can be calculated for each phase. This procedure is actually similar to that for synchronous generators:

$$\dot{E} = \dot{V} + \Delta V + j\delta V \qquad (6.47)$$

$$\Delta V = \frac{PR + QX}{V} \qquad (6.48)$$

$$\delta V = \frac{PX - QR}{V} \qquad (6.49)$$

where
X represents the filter reactance in the inverter
R represents the filter resistance in the inverter

In practice, the output power of the inverter may fluctuate during the first 1–2 cycles after a fault. However, because of the low inertia of the invert-ers and the fast adjustment of the controllers, the output power (both active and reactive) rapidly reaches its steady value, which is the same as the value before fault occurs. Besides, the large current that resulted from the external short-circuit may cause damage to the electronic devices that have limited ability to withstand over-current. For these reasons, the con-trollers set an upper limit for the output current of the inverter, generally about 1.5–2 times of the rated current.

Therefore, the power flowing from the inverter to the distribution system may also be assumed as constant before and after fault occurs. This output power can be calculated based on the power flow result before fault. The output current of the inverter is considered as constant if it reaches the upper limit.

6.7 Case Studies

6.7.1 Case for Power Flow Analysis

This case takes the modified IEEE 123-node test network as an example to demonstrate the procedure of power flow analysis.

Figure 6.25 shows the structure of the IEEE 123-node example and the parameters of this example can be obtained from the IEEE standard [18]. Three PV systems and one battery are connected to the system with the parameters shown in Table 6.5. Besides, the tie switch between

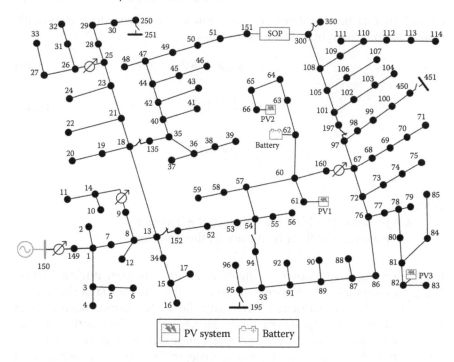

FIGURE 6.25 The IEEE 123-node network.

TABLE 6.5 Configuration of DGs

DG	Location	Real Power (kW)	Node Type
PV1	61	200	*PQ*
PV2	66	200	*PQ*
PV3	82	50	*PQ*
Battery	62	150	*PV*

node 151 and node 300 is replaced by an SOP, which controls the real power injected into node 300 (50 kW). Assume the reactive power of the two VSCs is zero. The power flow calculation result is given in Table 6.6.

It can been seen from Table 6.6 that the voltage profile of the distribution network is improved with DGs. In addition, the SOP controls the real power exchange to optimize the

TABLE 6.6 Power Flow Calculation Results

Node Number	Voltage Magnitude of Phase a (pu)	Voltage Magnitude of Phase b (pu)	Voltage Magnitude of Phase c (pu)
1	0.99244	0.99874	0.99491
13	0.97897	0.99599	0.98735
25	0.9883	0.99227	0.97615
35	0.96773	0.90039	0.97824
51	0.96254	0.98300	0.97333
67	0.97703	0.99152	0.98690
76	0.97536	0.98784	0.98548
87	0.97375	0.98514	0.98739
94	0.97241	0.98453	0.98798
95	0.97274	0.98397	0.98823
101	0.97552	0.99059	0.98563
108	0.97306	0.99094	0.98581
150	1.00000	1.00000	1.00000
160	0.98020	0.99427	0.99038
450	0.97640	0.98977	0.98532

power flow of the distribution network. The Zbus Gauss method shows good convergence and computational efficiency, which can be applied to power flow analyses of large-scale distribution networks.

6.7.2 Case for Sequential Power Flow

In this case, a 4-node distribution network is adopted to validate the procedure of sequential power flow.

A DG fluctuation can be smoothed using the ESS. As seen in Figure 6.26, whose case data are the same as in Example 6.7 excepting the PV system and ESS integrated at bus B2. Here, the PV output changes rapidly due to the cloud sheltering from sunshine. The controller of the ESS limits the active power of line 2 into a band of −1000 to 500 kW, whose control result for 1 h is validated in Figure 6.27.

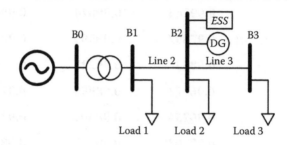

FIGURE 6.26 Case study of a 4-node distribution network with an ESS.

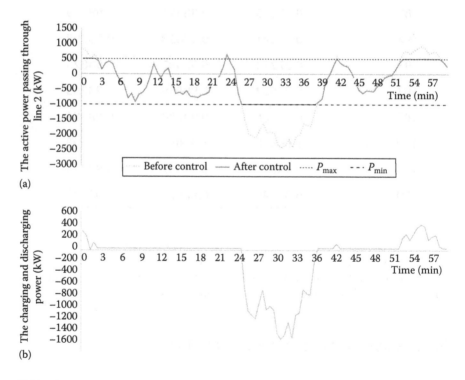

FIGURE 6.27 Control of the ESS: (a) active power passing through line 2 and (b) charging and discharging power.

As shown in Figure 6.27, the active power passing through line 2 violates the set band (−1000 to 500 kW) without controlling the ESS. When the active power of line 2 exceeds the upper limit (500 kW) caused by the excessive DG output, the ESS is controlled to charge. And the ESS is controlled to discharge if the active power of line 2 exceeds the lower limit (−1000 kW) caused by overload. Thus the active power passing through line 2 is limited into the set band by controlling the ESS.

6.7.3 Case for Short-Circuit Fault Analysis

This case takes a 4-node distribution network as an example to show the procedure of short-circuit fault analysis.

The DG connected to the distribution network shown in Figure 6.26 is assumed to be a PV station, which consists of eight PV units. For each PV unit, the rated capacity of the inverter is 0.5 MVA; the upper current limit is 1.5 times the rated current; the output power of the inverter is assumed constant before and after the fault. Now, assume that a three-phase to earth non-bolted fault occurs at node K, where the fault impedance is after the star point.

The fault reactance matrix derived in Example 6.8 is utilized to perform short-circuit calculation in this test case. The active and reactive power output of the PV inverter have been obtained in Example 6.4:

$$P_{inv} = 3.974 \text{ MW}, \quad Q_{inv} = 0 \text{ Mvar}$$

Based on the rated capacity, the rated current I_N of each PV inverter is calculated:

$$I_N = \frac{S_N}{\sqrt{3}V_N} = \frac{500 \times 8}{\sqrt{3} \times 10.5} = 219.94 \text{ A}$$

Then the upper current limit I_S is derived:

$$I_S = 1.5I_N = 329.91 \text{ A}$$

The calculation procedure is as follows:

1. Maintain the output power of the inverter P_{inv} as a constant 3.974 MW before and after the fault occurs.

2. Then the methodology of fault analysis presented in Section 6.6.2 is executed. By simulating the fault and solving the new power flow

equation, the RMS voltages of bus B0–B3 are calculated as 63.50853, 5.02401, 2.93279 and 0 kV. The output current for each phase of the inverter is 451.68 A, exceeding its upper limit.

3. Thus, the short-circuit current for each phase of the inverter is limited to 329.91 A, and the corresponding RMS voltages of bus B0–B3 are 63.50853, 4.99172, 2.81848 and 0 kV, respectively.

The output power of each PV inverter is calculated:

$$P_{\mathrm{inv,act}} = 3 \times 2.81848 \times 329.91 = 2.789 \text{ MW}$$

It can be seen that the PV inverter no longer keeps constant power output considering its upper current limit. Its output power is actually decreased to guarantee operational safety under the fault condition.

In conclusion, the short-circuit current at the fault point is derived by adding up the current from the external source (transformer substation) and the PV station; the final short-circuit current at the fault point is 2.062 kA.

☐ Questions

1. Explain the benefits of the application of SOPs in a distribution network.

2. Explain how to model distributed generation in power flow analysis.

3. The parameters of a 1 km long 10 kV overhead line are as follows: self-impedance $z_{aa} = (0.4576 + j1.0780)$ Ω/km, $z_{bb} = (0.4666 + j1.0482)$ Ω/km, $z_{cc} = (0.4615 + j1.0651)$ Ω/km; mutual impedance $z_{ab} = z_{ba} = (0.1560 + j0.5017)$ Ω/km, $z_{ac} = z_{ca} = (0.1535 + j0.3849)$ Ω/km, $z_{bc} = z_{cb} = (0.1580 + j0.4236)$ Ω/km; shunt capacitance $b_{aa} = 5.6765$ μS/km, $b_{bb} = 5.9809$ μS/km, $b_{cc} = 5.3971$ μS/km, $b_{ab} = -1.8319$ μS/km, $b_{ac} = -0.6982$ μS/km, $b_{bc} = -1.1645$ μS/km.

What is the admittance matrix of this line?

☐ References

1. EPRI US, *Electric Power Research Institute 2008 Portfolio: P161 IntelliGrid*. Palo Alto, CA, 2008.
2. S. Lee, A. Valenti and I. Bel, *Distribution Fast Simulation and Modeling (DFSM) High Level Requirements*. Palo Alto, CA: EPRI, 2005.

3. S. Wang and C. Wang, *Modern Distribution System Analysis*. Beijing, China: Higher Education Press, 2007.

4. W. Kersting, *Distribution System Modeling and Analysis*. Boca Raton, FL: CRC Press, 2012.

5. J. Arrillaga and N. Watson, *Computer Modeling of Electrical Power Systems*, 2nd edn. John Wiley & Sons Ltd., Chichester, U.K., 2001.

6. M. G. Molina and P. E. Mercado, Modeling and control of grid-connected photovoltaic energy conversion system used as a dispersed generator, *The 2008 IEEE/PES Transmission & Distribution Conference & Exposition*, Latin America, Bogota, Colombia, 2008, pp. 1–8.

7. Y. Levron, J. M. Guerrero and Y. Beck, Optimal power flow in microgrids with energy storage, *IEEE Transactions on Power Systems*, 28(3), 3226–3234, 2013.

8. S. Arya and B. Singh, Implementation of kernel incremental metalearning algorithm in distribution static compensator, *IEEE Transactions on Power Electronics*, 30(3), 1157–1169, 2015.

9. J. Nielsen, M. Newman, H. Nielsen et al., Control and testing of a dynamic voltage restorer (DVR) at medium voltage level, *IEEE Transactions on Power Electronics*, 19(3), 806–813, 2004.

10. J. M. Bloemink and T. C. Green, Increasing photovoltaic penetration with local energy storage and soft normally-open points, *The 2011 IEEE Power and Energy Society General Meeting*, San Diego, CA, 24–29 July 2011, pp. 1–8.

11. J. M. Bloemink and T. C. Green, Benefits of distribution-level power electronics for supporting distributed generation growth, *IEEE Transactions on Power Delivery*, 28(2), 911–919, 2013.

12. J. He, B. Zhou, Q. Zhang et al., An improved power flow algorithm for distribution networks based on zbus algorithm and forward/backward sweep method, *Proceedings of 2012 International Conference on Control Engineering and Communication Technology*, Liaoning, China, 2012, pp. 1–4.

13. F. Zhang and C. S. Cheng, A modified newton method for radial distribution system power flow analysis, *IEEE Transactions on Power Systems*, 12(1), 389–397, 1997.

14. EPRI, Introduction to the OpenDSS, 2009. Available from: http://electricdss. svn.sourceforge.net/, accessed on September 11, 2016.

15. K. P. Schneide and J. C. Fuller, Voltage control devices on the IEEE 8500 node test feeder, *The 2010 IEEE/PES Transmission & Distribution Conference & Exposition*, New Orleans, LA, 19–22 April 2010, pp. 1–6.

16. C. Wang and X. Sun, The improved short circuit calculation method of distributed network containing distributed generation, *Automation of Electric Power Systems*, 36(23), 54–58, 2012.

17. N. Natthaphob, Fault current contribution from synchronous machine and inverter based distributed generators, *IEEE Transactions on Power Delivery*, 22(1), 631–641, 2007.

18. IEEE PES Distribution System Analysis Subcommittee, Distribution Test Feeders. Available from: http://ewh.ieee.org/soc/pes/dsacom/testfeeders/ index.html, accessed on September 11, 2016.

CHAPTER

7

Transient Analysis

7.1 Introduction

With the ability to connect renewable energy generation and distributed energy storage and the advent of novel power electronic equipment, distribution systems are changing from passive networks to active networks. It is important to study the transient behaviour of such active distribution networks as stability and reliability may be compromised due to the variation of renewable generation and switching of power electronic equipment. This chapter addresses the transient analysis of distribution systems through electromagnetic transient simulations.

7.2 First Example of Electromagnetic Transient Simulation

In order to introduce the problem formulation and basic concepts of the electromagnetic transient simulation, the boost circuit shown in Figure 7.1 is used as an example.

The inductor current i_L and the capacitor voltage v_C are selected as the state variable, i.e. $x = [i_L, v_C]^T$. Assuming ideal switch characteristics

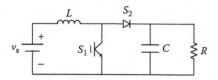

FIGURE 7.1 The boost circuit.

of the two power electronic devices S_1 and S_2, the differential equation describing the boost circuit can be established:

$$px = Ax + Bv_s \qquad (7.1)$$

where
    p is the differential operator
    A and B are the state matrix and input matrix respectively, whose values depend on the status of S_1 and S_2

When S_1 is on and S_2 is off, the boost circuit works in mode 1. Since S_2 is off, the left-hand side and the right-hand side of the circuit can be considered separately. Then the following equations can be written:

$$L\frac{di_L}{dt} = v_s \qquad (7.2)$$

$$C\frac{dv_C}{dt} - \frac{v_C}{R} = 0 \qquad (7.3)$$

These equations can be written in matrix form as

$$p\begin{bmatrix} i_L \\ v_C \end{bmatrix} = \begin{bmatrix} 0 & 0 \\ 0 & -\frac{1}{RC} \end{bmatrix}\begin{bmatrix} i_L \\ v_C \end{bmatrix} + \begin{bmatrix} \frac{1}{L} \\ 0 \end{bmatrix}v_s \qquad (7.4)$$

Comparing this equation with Equation 7.1, we have

$$A = \begin{bmatrix} 0 & 0 \\ 0 & -\frac{1}{RC} \end{bmatrix}, \quad B = \begin{bmatrix} \frac{1}{L} \\ 0 \end{bmatrix} \qquad (7.5)$$

When S_1 is off and S_2 is on, the boost circuit works in mode 2.

$$v_s = L\frac{di_L}{dt} + v_C \qquad (7.6)$$

$$i_L + C \frac{dv_C}{dt} - \frac{v_C}{R} = 0 \tag{7.7}$$

These equations can be written in matrix form as

$$p \begin{bmatrix} i_L \\ v_C \end{bmatrix} = \begin{bmatrix} 0 & -\dfrac{1}{L} \\ \dfrac{1}{C} & -\dfrac{1}{RC} \end{bmatrix} \begin{bmatrix} i_L \\ v_C \end{bmatrix} + \begin{bmatrix} \dfrac{1}{L} \\ 0 \end{bmatrix} v_s \tag{7.8}$$

Comparing this equation with Equation 7.1, we have

$$A = \begin{bmatrix} 0 & -\dfrac{1}{L} \\ \dfrac{1}{C} & -\dfrac{1}{RC} \end{bmatrix}, \quad B = \begin{bmatrix} \dfrac{1}{L} \\ 0 \end{bmatrix} \tag{7.9}$$

With the differential equation (7.1), the simulation curves for the state variable $\{x(t)| t \in [0, T_{end}], x(0) = x_0\}$ can be formed, where x_0 is the initial value of the state variable for the simulation and T_{end} is the time duration of simulation. The solution is obtained by inserting N discretization nodes $\{t_n | n = 1, 2, ..., N, \text{ s.t. } 0 < t_1 < t_2 < \cdots < T_{end}\}$ into the interval $[0, T_{end}]$ and dividing this interval into $N + 1$ smaller intervals $[t_n, t_{n+1}]$, which are called integration steps. Numerical integration formulas are then applied in these small steps. Let x_n be the state variable at t_n and x_{n+1} be the state variable at t_{n+1}. The simulation step size Δt is defined as $\Delta t = t_{n+1} - t_n$. We apply the widely used implicit trapezoidal formula on Equation 7.1 and derive the following algebraic equation:

$$\left(I - \frac{\Delta t}{2} A \right) x_{n+1} = \left(I + \frac{\Delta t}{2} A \right) x_n + \frac{\Delta t}{2} B \left(v_s(t_{n+1}) + v_s(t_n) \right) \tag{7.10}$$

Using Equation 7.10 iteratively, the state variables at discretization nodes $x_1, x_2, ...$ can be computed successively, starting from the initial value x_0.

It is worth mentioning that if any power electronic device, e.g. S_1 or S_2, switches at a particular simulation step, additional interpolation computation is needed to find the exact switch time. Interpolation is important for the simulation accuracy of distribution systems where power electronic devices are pervasive. Section 7.5 will delve into more details about interpolation, but in this simple example, we will assume all the switching instants are on the discretization nodes $\{t_n\}$.

Let us consider a fixed duty ratio control of the boost circuit and set the initial values to zero. The numerical simulation is performed as explained earlier to simulate the starting dynamics of the boost circuit. The simulation curves are shown in Figure 7.2.

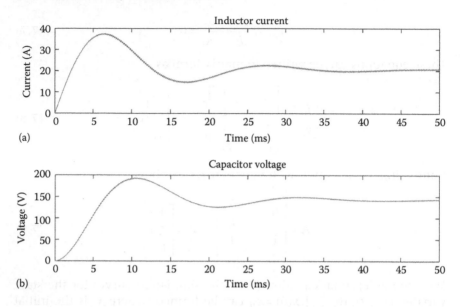

FIGURE 7.2 Electromagnetic transient simulation of the boost circuit: (a) the inductor current wave form and (b) the capacitor voltage wave form.

Figure 7.2 shows the dynamics of the inductor current and capacitor voltage in the boost circuit that ramp from zero to steady state. From the simulation curves, the performance of the circuit such as the ramp-up time, overshoot ratio, and the steady-state ripples on the load can be quantitatively measured. This information can be further utilized to inform the parameter design of the circuit. The MATLAB® code used in this example is provided in Section 7.A.1. Readers are encouraged to run the code, make adjust the parameters and observe their influence on the simulation curve.

Although the modelling and simulation in this section are focused on a simple boost circuit, this demonstration of the modeling process is applicable for general power systems.

7.3 Nodal Analysis for Transient Simulation

For electromagnetic transient simulations, two approaches are used, i.e. nodal analysis and state space analysis. State space analysis was used in the boost circuit example in Section 7.2. The basic process is to establish a set of differential equations (state equations) that describe the dynamics of the system and then apply a numerical integration formula to the differential equations as a whole. The major advantage of state space analysis is the flexibility in choosing the integration algorithms. However, the formation of the state equations for a large power system could be challenging. Nodal analysis, on the other hand, uses a bottom-up modelling approach that applies the integration formulas at the branch level, which significantly simplifies the modelling process. Many commercial electromagnetic transient simulation programs use the nodal analysis approach, and they are commonly referred to as (electromagnetic transients programme)-type programmes [1,2]. The following sections focus on the nodal analysis approach.

7.3.1 Electrical System Element Modelling

A numerical integration formula is used to discretize the differential equations corresponding to each element into algebraic equations. These algebraic equations are frequently represented in the form of equivalent circuits of the elements, referred to as the companion circuit model of the element in the literature. The ensemble circuit formed by connecting the equivalent circuits of each element is then analysed using the nodal analysis method and a set of algebraic equations are derived. These equations are then solved using numerical methods.

The implicit trapezoidal method is the most commonly used integration formula in the nodal analysis framework. The discretization rule can be summarized as follows:

1. Replace variable px with $(x(t) - x(t - \Delta t))/\Delta t$, where p is the differential operator.

2. Replace variable x with $(x(t) + x(t - \Delta t))/2$.

3. Constants and constant coefficients remain unchanged.

Take the inductor branch shown in Figure 7.3a as an example; the *V-I* relationship of the branch is described by Equation 7.11.

$$L \cdot p i_{km} = v_k - v_m \tag{7.11}$$

Applying the discretization rule of the trapezoidal method in Equation 7.11, the resultant algebraic equation is

$$L \left[\frac{i_{km}(t) - i_{km}(t - \Delta t)}{\Delta t} \right] = \frac{v_k(t) + v_k(t - \Delta t)}{2} - \frac{v_m(t) + v_m(t - \Delta t)}{2} \tag{7.12}$$

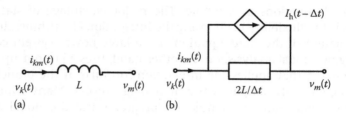

$i_{km}(t)$

$v_k(t)$ L $v_m(t)$

(a)

$i_{km}(t)$

$v_k(t)$ $2L/\Delta t$ $v_m(t)$

$I_h(t-\Delta t)$

(b)

FIGURE 7.3 Inductor branch and its equivalent circuit: (a) inductor branch and (b) equivalent circuit.

This equation can be rearranged as

$$i_{km}(t) = \frac{\Delta t}{2L}\left(v_k(t) - v_m(t)\right) + I_h(t-\Delta t) \qquad (7.13)$$

where $I_h(t-\Delta t)$ is referred to as the history current source, with the form

$$I_h(t-\Delta t) = i_{km}(t-\Delta t) + \frac{\Delta t}{2L}\left(v_k(t-\Delta t) - v_m(t-\Delta t)\right) \qquad (7.14)$$

The difference equation (7.13) can be viewed as a Norton equivalent circuit with an equivalent conductance of $G_{eq} = \Delta t/2L$ in parallel with the history current source, as shown in Figure 7.3b.

Like the inductor element, most fundamental elements of power system models can be treated in a similar way using the discretization rule. For the common linear circuit elements, i.e. resistors, inductors, capacitors (abbreviated as RLC hereafter), the expressions of the equivalent conductance G_{eq} and the history current source I_h are summarized in Table 7.1.

Among the multi-phase elements with mutual coupling, the coupled inductor element shown in Figure 7.4 is most frequently seen. Using the familiar implicit trapezoidal method as the discretization rule, the characteristic equation of the element, equivalent conductance and history current source for the multi-phase elements can be expressed in a similar way to the single-phase elements in matrix–vector notation:

$$L \cdot pi_{km} = v_k - v_m \qquad (7.15)$$

$$G_{eq} = \frac{\Delta t L^{-1}}{2} \qquad (7.16)$$

$$I_h = i_{km}(t-\Delta t) + \frac{\Delta t L^{-1}\left[v_k(t-\Delta t) - v_m(t-\Delta t)\right]}{2} \qquad (7.17)$$

TABLE 7.1 Conductance and History Current Source of RLC Elements

Element		Discretization Using the Trapezoidal Rule
Resistor R	Original equation	$v_k - v_m = R \cdot i_{km}$
	Difference equation	$i_{km}(t) = \dfrac{1}{R}\left(v_k(t) - v_m(t)\right)$
	G_{eq}	$1/R$
	I_h	0
Inductor L	Original equation	$v_k - v_m = L \cdot p i_{km}$
	Difference equation	$i_{km}(t) = \dfrac{\Delta t}{2L}\left[v_k(t) - v_m(t)\right] + I_h(t - \Delta t)$
	G_{eq}	$\Delta t/(2L)$
	I_h	$i_{km}(t - \Delta t) + \dfrac{\Delta t}{2L}\left[v_k(t - \Delta t) - v_m(t - \Delta t)\right]$

(Continued)

TABLE 7.1 (*Continued*) Conductance and History Current Source of RLC Elements

Element		Original equation	Discretization Using the Trapezoidal Rule
Capacitor C	Original equation	$C \cdot p(v_k - v_m) = i_{km}$	
	Difference equation	$i_{km}(t) = \dfrac{2C}{\Delta t}\left[v_k(t) - v_m(t)\right] + I_h(t - \Delta t)$	
	G_{eq}	$2C/\Delta t$	
	I_h	$-i_{km}(t - \Delta t) - \dfrac{2C}{\Delta t}\left[v_k(t - \Delta t) - v_m(t - \Delta t)\right]$	

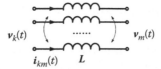

$v_k(t)$ $\qquad$ $v_m(t)$

$i_{km}(t)$ $\quad$ L

FIGURE 7.4 Multi-phase coupled inductor branch.

Discretization at the element level requires the numerical integration formula to have a simple form, so that the expression of the equivalent conductance and history current source is not too complicated, and with good numerical stability and accuracy. The implicit trapezoidal formula is usually a good choice for the electromagnetic transient simulation.

With an understanding of the discretization process on the element level, models of more complicated systems can be established, as illustrated in Example 7.1.

Example 7.1

Convert the boost circuit shown in Figure 7.1 to an equivalent nodal representation. The power electronic switches are assumed to have ideal characteristics.

Answer

Using the discretize RLC elements given in Table 7.1, the boost circuit is represented by the equivalent circuit model and is shown in Figure 7.5.

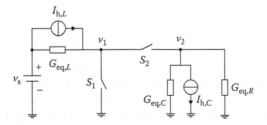

FIGURE 7.5 Equivalent circuit model of the boost circuit.

7.3.2 Modelling and Simulation at the System Level

After assembling an equivalent circuit comprised of voltage sources, current sources, conductance and switches (e.g. Figure 7.5), nodal analysis is applied to establish a set of equations that represent the nodal voltage and branch current constraint, alongside the equations of each element. This set of equations is the fundamental equation of the electromagnetic transient simulation of distribution systems and is in the form of Equation 7.18:

$$Gv = i \qquad (7.18)$$

where the G matrix and i vector consist of the equivalent conductance G_{eq} and the history current source I_h. For illustration, the contribution to the G matrix and i vector from a single-phase element that is connected between node k and m is shown in Figure 7.6a, and the contribution from a three-phase element connected between $\{k_1, k_2, k_3\}$ and $\{m_1, m_2, m_3\}$ is shown in Figure 7.6b.

The solution of Equation 7.18 gives the transient response of the system at the specified time instant. As Equation 7.18 is in the form of a linear system $Ax = b$, a sparse library such as Sparse [3], KLU [4], etc. can be used to solve it.

Modelling with the nodal analysis formulation has certain limitations, such as the inconvenience in handling current-dependent elements (e.g. the ideal transformer model) and ungrounded ideal voltage sources. The electronic circuit simulation program SPICE uses the modified nodal approach (MNA) to address these issues [5]. Modified augmented nodal analysis (MANA) further enhances the handling of ideal switch branches [6]. The fundamental equation under the MANA framework is shown as follows:

$$
\begin{bmatrix}
\mathbf{Y}_n & \mathbf{V}_c & \mathbf{D}_c & \mathbf{S}_c \\
\mathbf{V}_r & \mathbf{V}_d & \mathbf{D}_{VD} & \mathbf{S}_{VS} \\
\mathbf{D}_r & \mathbf{D}_{DV} & \mathbf{D}_d & \mathbf{S}_{DS} \\
\mathbf{S}_r & \mathbf{S}_{SV} & \mathbf{S}_{SD} & \mathbf{S}_d
\end{bmatrix}
\begin{bmatrix}
\mathbf{v}_n \\
\mathbf{i}_V \\
\mathbf{i}_D \\
\mathbf{i}_S
\end{bmatrix}
=
\begin{bmatrix}
\mathbf{i}_n \\
\mathbf{v}_b \\
\mathbf{d}_b \\
\mathbf{s}_b
\end{bmatrix}
\qquad (7.19)
$$

The MANA formulation (7.19) includes the voltage source branch current $\mathbf{i}_V$, the controlled source branch current $\mathbf{i}_D$, and the ideal switch branch current $\mathbf{i}_S$ as unknown variables in addition to the nodal voltage $\mathbf{v}_n$. The device equations of these branches are then listed below the nodal equations. For a voltage source connected between node k and m, the branch equation is $v_k - v_m = v_{b,km}$, which is inserted into Equation 7.19 by adding 1 and -1 in column k and m, respectively, of matrix $\mathbf{V}_r$ and including $v_{b,km}$ in the vector $\mathbf{v}_b$. Take an ideal switch connected between node p and q as another example: when the switch is on, equation $v_p - v_q = 0$ is included, which again contributes 1 and -1 to matrix $\mathbf{S}_r$; when the switch is off,

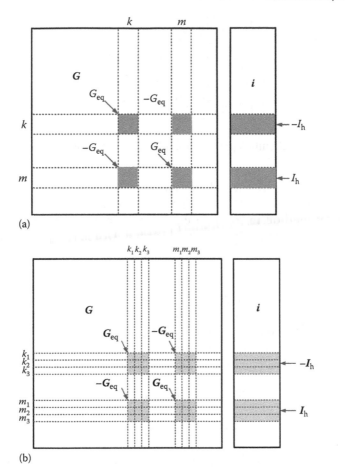

FIGURE 7.6 Formation of the nodal conductance matrix and current source vector: (a) single-phase element and (b) multi-phase coupled element.

the equation $i_{S,pq} = 0$ is included, which contributes 1 in the corresponding column of $\mathbf{S}_d$. In this chapter, the MANA method is used in modelling electrical distribution systems. Example 7.2 demonstrates this process using a typical distributed generation system.

Example 7.2

Transient modelling of a PV generation system

The circuit diagram of a grid-connected single-phase PV generation system is shown in Figure 7.7, where the PV module

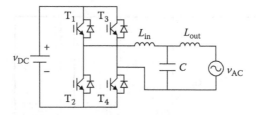

FIGURE 7.7 Circuit diagram of a single-phase grid-connected PV system.

is represented by the voltage source v_{DC} and the external grid is represented by v_{AC}. Use the discretization rule of the implicit trapezoidal method to form an equivalent circuit, and establish the MANA equations. The power electronic switches are assumed to be ideal.

Answer

The solution of a circuit needs a reference zero voltage. We select the mid-point of the DC voltage source v_{DC} as zero. Using the implicit trapezoidal method, the computational equivalent circuit shown in Figure 7.8 is established.

The nodal equations:

$$\text{Node 1:} \quad -i_{DC_1} + i_{T_1} + i_{T_3} = 0 \tag{7.20}$$

$$\text{Node 2:} \quad i_{DC_2} - i_{T_2} - i_{T_4} = 0 \tag{7.21}$$

$$\text{Node 3:} \quad G_{L_{in}}(v_3 - v_5) - i_{T_1} + i_{T_2} = -I_{L_{in}} \tag{7.22}$$

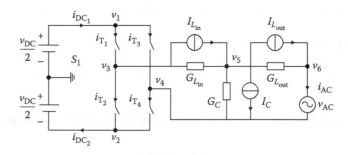

FIGURE 7.8 Computational equivalent circuit of the grid-connected PV system.

Node 4: $G_C\left(v_4 - v_5\right) - i_{\text{AC}} - i_{\text{T}_3} + i_{\text{T}_4} = I_C$ (7.23)

Node 5: $G_{L_{\text{in}}}\left(v_5 - v_3\right) + G_C\left(v_5 - v_4\right)$
$+ G_{L_{\text{out}}}\left(v_5 - v_6\right) = I_{L_{\text{in}}} - I_C - I_{L_{\text{out}}}$ (7.24)

Node 6: $G_{L_{\text{out}}}\left(v_6 - v_5\right) + i_{\text{AC}} = I_{L_{\text{out}}}$ (7.25)

The voltage source branches:

DC source 1: $v_1 = \dfrac{v_{\text{DC}}}{2}$ (7.26)

DC source 2: $v_2 = \dfrac{-v_{\text{DC}}}{2}$ (7.27)

AC source: $v_6 - v_4 = v_{\text{AC}}$ (7.28)

The ideal switch branches:

if T_1 is on: $v_1 - v_3 = 0$ (7.29)

if T_2 is on: $v_2 - v_3 = 0$ (7.30)

if T_3 is on: $v_1 - v_4 = 0$ (7.31)

if T_4 is on: $v_2 - v_4 = 0$ (7.32)

if T_k is off: $i_{\text{T}_k} = 0,\quad k = \{1,2,3,4\}$ (7.33)

Combining equations from (7.20) through (7.23), the matrix–vector equation $\boldsymbol{Gv} = \boldsymbol{i}$ is established, where

$$\boldsymbol{v} = \left[v_1, v_2, v_3, v_4, v_5, v_6, i_{\text{DC}_1}, i_{\text{DC}_2}, i_{\text{AC}}, i_{\text{T}_1}, i_{\text{T}_2}, i_{\text{T}_3}, i_{\text{T}_4}\right]^{\text{T}}$$ (7.34)

$$\boldsymbol{i} = \left[0,\ 0,\ -I_{L_{\text{in}}},\ I_C,\ I_{L_{\text{in}}} - I_C - I_{L_{\text{out}}},\ I_{L_{\text{out}}},\ \frac{v_{\text{DC}}}{2},\ \frac{-v_{\text{DC}}}{2},\ v_{\text{AC}},\ 0,\ 0,\ 0,\ 0\right]^{\text{T}}$$ (7.35)

The elements of G depend on the on/off status of the switches $\{T_1, T_2, T_3, T_4\}$, with totally $2^4 = 16$ different combinations (not all the combinations occur in practice, e.g. both T_1 and T_2 are not on simultaneously). We list only the situation when T_1 and T_4 are on and T_2 and T_3 are off:

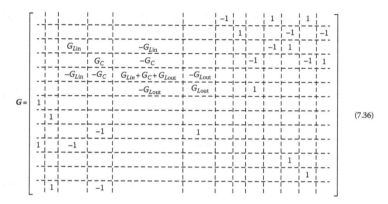

$$(7.36)$$

In Equation 7.36, the blank spots of the G matrix are zeros.

7.4 Control System Simulation in the Nodal Analysis Framework

The building blocks of a control system include general transfer functions, basic math and logic operations, various non-linear and discontinuity blocks, etc. Unlike electrical systems, the characteristics of control system blocks are described using their input–output functions, which are typically non-linear. Therefore, the solution of control systems requires an approach different from modelling of power system components. This section introduces the compositional modelling method of control systems and the solution of the resultant equations using Newton's method.

7.4.1 Compositional Modelling of Control Systems

The first step of control system modelling is to write down the input–output equations of all the blocks of the control system. Taking Figure 7.9 as an example, the input–output equations of the control system include:

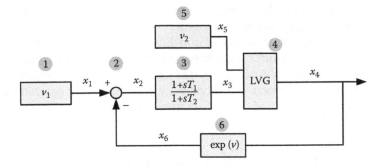

FIGURE 7.9 A control system example.

1. Signal source: $x_1 = v_1$ (7.37)

2. Summation: $x_2 = x_1 - x_6$ (7.38)

3. Lead-lag: $x_3(t) = k_1 x_2(t) + c_3(t - \Delta t)$ (7.39)

4. Low-value gate (LVG): $x_4 = x_3$ or $x_4 = x_5$ (7.40)

5. Signal source: $x_5 = v_2$ (7.41)

6. Exponential: $x_6 = \exp(x_4)$ (7.42)

Among them, the lead-lag Equation 7.39 is established using numerical integration methods, where the c_3 term is analogous to the history current source I_h in the discretization of electrical systems. The discretization of dynamic components in control systems, such as transfer functions, will be presented later.

Assuming the low-value gate has selected the lower channel as the output, i.e. $x_4 = x_3$, the complete set of equations is as follows:

$$
\begin{bmatrix}
1 & 0 & 0 & 0 & 0 & 0 \\
-1 & 1 & 0 & 0 & 0 & 1 \\
0 & -k_1 & 1 & 0 & 0 & 0 \\
0 & 0 & -1 & 1 & 0 & 0 \\
0 & 0 & 0 & 0 & 1 & 0 \\
0 & 0 & 0 & 0 & 0 & 1
\end{bmatrix}
\begin{bmatrix}
x_1(t) \\
x_2(t) \\
x_3(t) \\
x_4(t) \\
x_5(t) \\
x_6(t)
\end{bmatrix}
=
\begin{bmatrix}
v_1(t) \\
0 \\
c_3(t - \Delta t) \\
0 \\
v_2(t) \\
\exp(x_4(t))
\end{bmatrix}
\quad (7.43)
$$

In the modelling process, the number of output variables equals the number of input–output equations. The complete control system model is obtained by listing all the input–output equations of the control system blocks, with the number of equations being equal to the number of unknowns. Due to the existence of non-linearity (the exponential block), the model (7.43) requires non-linear system solution algorithms. All controllers in the distribution networks can be modelled using the demonstrated approach, where the models are established by combining the input–output equations of the basic control elements.

7.4.2 Modelling Dynamic Components in Control Systems

Besides the static elements in the control system, usually there are also dynamic elements in the form of transfer functions. These elements can be handled by discretizing using numerical integration methods, and then combining the resultant algebraic equations with the input–output equations of other elements. Without losing generality, we demonstrate the process with a transfer function of order n.

The general form of a transfer function of order n is

$$X(s) = k\frac{N_0 + N_1 s + \cdots + N_m s^m}{D_0 + D_1 s + \cdots + D_n s^n} V(s) \tag{7.44}$$

where
   V is the input signal
   X is the output signal

The differential equation form is

$$k\left(N_0 v + N_1 pv + \cdots + N_m p^m v\right) = D_0 x + D_1 px + \cdots + D_n p^n x \tag{7.45}$$

where p is the differential operator.

With the introduction of the auxiliary variables

$$x_1 = px, x_2 = px_1, \ldots, x_n = px_{n-1}$$
$$v_1 = pv, v_2 = pv_1, \ldots, v_m = pv_{m-1} \tag{7.46}$$

Equation 7.45 is converted into an algebraic equation:

$$k\left(N_0 v + N_1 v_1 + \cdots + N_m v_m\right) = D_0 x + D_1 x_1 + \cdots + D_n x_n \tag{7.47}$$

Applying the trapezoidal discretization rule to each equation in Equation 7.46 yields

$$x_i(t) = \frac{2}{\Delta t} x_{i-1}(t) - \left[x_i(t - \Delta t) + \frac{2}{\Delta t} x_{i-1}(t - \Delta t) \right]$$

$$v_i(t) = \frac{2}{\Delta t} v_{i-1}(t) - \left[v_i(t - \Delta t) + \frac{2}{\Delta t} v_{i-1}(t - \Delta t) \right]$$

(7.48)

where $x_0 = x$, $v_0 = v$. Substituting (7.48) into (7.47) eliminates all the auxiliary variables. With some algebraic manipulation, the following simple linear equation is derived:

$$c_0 x(t) = k d_0 v(t) + h_1(t - \Delta t)$$

(7.49)

where the history term h_1 in each step depends on the history term in the previous step, with the following recursive relation:

$$h_1(t) = k d_1 v(t) - c_1 x(t) - h_1(t - \Delta t) + h_2(t - \Delta t)$$

......

$$h_i(t) = k d_i v(t) - c_i x(t) - h_i(t - \Delta t) + h_{i+1}(t - \Delta t)$$

(7.50)

......

$$h_n(t) = k d_n u(t) - c_n x(t)$$

The values of c_i and d_i are also derived in a recursive manner. Due to the similarity of their form, only the formula of c_i is given here:

$$c_0 = \sum_{i=0}^{n} \left(\frac{2}{\Delta t} \right)^i D_i$$

$$c_i = c_{i-1} + (-2)^i \sum_{j=i}^{n} c_j^i \left(\frac{2}{\Delta t} \right)^j D_j$$

(7.51)

The computation formula for d_i is easily obtained by replacing D_i with N_i and n by m. The 0th order transfer function, i.e. the common gain element, is a special case of the general transfer function, where the history term equals 0 and $c = d = 1$.

Example 7.3

In Section 7.4.1, the input-output equation of the lead-lag element

$$x(t) = k u(t) + c(t - \Delta t)$$

(7.52)

is used with its derivation deferred. Using the method described in this section, derive this input-output equation, and specify the expression of the coefficient k and the history term $c(t - \Delta t)$.

Answer

The transfer function of lead-lag is

$$X(s) = \frac{1 + sT_1}{1 + sT_2} V(s) \tag{7.53}$$

Comparing it with the general form of the nth order transfer function (7.44), the following correspondence is seen:

$$k = 1 \tag{7.54}$$

$$N_0 = 1, \quad N_1 = T_1, \quad m = 1 \tag{7.55}$$

$$D_0 = 1, \quad D_1 = T_2, \quad n = 1 \tag{7.56}$$

Thus, we can directly use the result for the general transfer functions (7.49) and (7.50) and get the input–output equation:

$$c_0 x(t) = d_0 v(t) + h_1(t - \Delta t) \tag{7.57}$$

$$h_1(t - \Delta t) = d_1 v(t - \Delta t) - c_1 x(t - \Delta t) \tag{7.58}$$

where the coefficients are

$$c_0 = D_0 + 2D_1/\Delta t = 1 + 2T_2/\Delta t \tag{7.59}$$

$$c_1 = c_0 + (-2) \cdot 2D_1/\Delta t = 1 - 2T_2/\Delta t \tag{7.60}$$

$$d_0 = 1 + 2T_1/\Delta t \tag{7.61}$$

$$d_1 = 1 - 2T_1/\Delta t \tag{7.62}$$

Hence, Equations 7.57 through 7.62 can be rewritten in the form required by the question, where

$$k = \frac{\Delta t + 2T_1}{\Delta t + 2T_2} \tag{7.63}$$

$$c(t - \Delta t) = \frac{\Delta t - 2T_1}{\Delta t + 2T_2} v(t - \Delta t) - \frac{\Delta t - 2T_2}{\Delta t + 2T_2} x(t - \Delta t) \tag{7.64}$$

7.4.3 Control System Transient Simulation Based on Newton's Method

As mentioned, the model of a control system is fundamentally different from the model of an electrical system. As the majority of the electrical system elements are linear, the electrical network is modelled as a set of linear equations, where the non-linearity is handled locally. As for a control system, the vast number of non-linear elements requires the system to be modelled as a set of non-linear equations and be solved using iterative methods.

An nth order non-linear system

$$\begin{cases} f_1(x_1, x_2,..., x_n) = 0 \\ f_2(x_1, x_2,..., x_n) = 0 \\ \quad\vdots \\ f_n(x_1, x_2,..., x_n) = 0 \end{cases} \tag{7.65}$$

can be written in vector form as

$$f(x) = 0 \tag{7.66}$$

where

$$f = [f_1, f_2, ..., f_n]^T$$
$$x = [x_1, x_2, ..., x_n]^T$$

Newton's method is popular for solving non-linear systems in a wide range of engineering applications. The iteration form of Newton's method is

$$f\left(x^{(k)}\right) + J^{(k)}\Delta x^{(k)} = 0 \tag{7.67}$$

As a fundamental equation of the control system, Equation 7.67 has a role similar to the equation $Gv = i$ for the electrical system introduced in Section 7.3, where $\Delta x^{(k)} = \left[\Delta x_1^{(k)}, \Delta x_2^{(k)},..., \Delta x_n^{(k)}\right]^T$ and $J^{(k)}$ is the Jacobian matrix:

$$J^{(k)} = \begin{bmatrix} \dfrac{\partial f_1\left(x_1^{(k)}, x_2^{(k)},..., x_n^{(k)}\right)}{\partial x_1} & \dfrac{\partial f_1\left(x_1^{(k)}, x_2^{(k)},..., x_n^{(k)}\right)}{\partial x_2} & \cdots & \dfrac{\partial f_1\left(x_1^{(k)}, x_2^{(k)},..., x_n^{(k)}\right)}{\partial x_n} \\[2em] \dfrac{\partial f_2\left(x_1^{(k)}, x_2^{(k)},..., x_n^{(k)}\right)}{\partial x_1} & \dfrac{\partial f_2\left(x_1^{(k)}, x_2^{(k)},..., x_n^{(k)}\right)}{\partial x_2} & \cdots & \dfrac{\partial f_2\left(x_1^{(k)}, x_2^{(k)},..., x_n^{(k)}\right)}{\partial x_n} \\[2em] \vdots & \vdots & \ddots & \vdots \\[2em] \dfrac{\partial f_n\left(x_1^{(k)}, x_2^{(k)},..., x_n^{(k)}\right)}{\partial x_1} & \dfrac{\partial f_n\left(x_1^{(k)}, x_2^{(k)},..., x_n^{(k)}\right)}{\partial x_2} & \cdots & \dfrac{\partial f_n\left(x_1^{(k)}, x_2^{(k)},..., x_n^{(k)}\right)}{\partial x_n} \end{bmatrix} \tag{7.68}$$

If the entire control system is linear, the Jacobian matrix is a constant matrix, and only one linear system needs to be solved in each simulation step. If there are non-linear elements in the system, the control system needs to be solved iteratively. Consider the example in Figure 7.9; its Jacobian matrix is

$$J^{(k)} = \begin{bmatrix} 1 & 0 & 0 & 0 & 0 & 0 \\ -1 & 1 & 0 & 0 & 0 & 1 \\ 0 & -k_1 & 1 & 0 & 0 & 0 \\ 0 & 0 & -1 & 1 & 0 & 0 \\ 0 & 0 & 0 & 0 & 1 & 0 \\ 0 & 0 & 0 & -\exp\left(x_4^{(k)}\right) & 0 & 1 \end{bmatrix} \tag{7.69}$$

The iteration process is as follows:

1. Set the initial value of the iteration $x^{(0)}$.

2. Compute the initial Jacobian matrix $J^{(0)}$ according to Equation 7.69, then solve the linear equation $J^{(0)}\Delta x^{(0)} = -f(x^{(0)})$.

3. Update $x^{(1)} = x^{(0)} + \Delta x^{(0)}$.

4. Compute the Jacobian matrix $J^{(1)}$ according to Equation 7.69, then solve the linear equation $J^{(1)}\Delta x^{(1)} = -f(x^{(1)})$.

5. Update $x^{(2)} = x^{(1)} + \Delta x^{(1)}$.

6. ...

The iteration continues until the desired precision is achieved.

Besides the continuous non-linearity such as the exponential function, control systems in distribution networks also contain discontinuous non-linearities (hard non-linearities). In this case, the equation describing the input–output relationship of the element is in the form of a piecewise function, and the solution requires occasionally reformulating the Jacobian matrix. This is similar to the handling of switch models in an electrical system and deserves careful attention.

Next, we demonstrate the modelling and simulation process of the control system using a typical control module.

Example 7.4

Figure 7.10 shows the diagram of a PLL system, which is a frequently encountered control system module in distributed generation sources and energy storage systems. The module is used to measure system frequency and phase angle and is composed of three components: phase detector, loop filter and voltage-controlled oscillator.

Derive the compositional model of the PLL system, and specify the iteration form of the Jacobian matrix.

Answer

1. The input–output equation of the phase detection component is

$$v_\alpha = \frac{2}{3}v_a - \frac{1}{3}v_b - \frac{1}{3}v_c \qquad (7.70)$$

$$v_\beta = \frac{1}{\sqrt{3}}v_b - \frac{1}{\sqrt{3}}v_c \qquad (7.71)$$

$$v_q = v_\alpha \cos\theta - v_\beta \sin\theta \qquad (7.72)$$

2. The loop filter component is a proportional–integral (PI) controller, which can be handled as a transfer function. The transfer function of the PI controller is

$$X(s) = \frac{K_I + sK_P}{s}V(s) \qquad (7.73)$$

Comparing it with the general form of the nth order transfer function (7.44), the following correspondence is observed:

$$k = 1 \qquad (7.74)$$

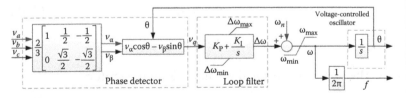

FIGURE 7.10 Diagram of a PLL system.

$$N_0 = K_I, \quad N_1 = K_P, \quad m = 1 \tag{7.75}$$

$$D_0 = 0, \quad D_1 = 1, \quad n = 1 \tag{7.76}$$

Using Equations 7.49 and 7.50, the input–output equation of the PI block is obtained:

$$c_0 x(t) = d_0 v(t) + h_1(t - \Delta t) \tag{7.77}$$

$$h_1(t - \Delta t) = d_1 v(t - \Delta t) - c_1 x(t - \Delta t) \tag{7.78}$$

where the coefficients are obtained using Equation 7.51:

$$c_0 = \frac{2}{\Delta t} \tag{7.79}$$

$$c_1 = \frac{-2}{\Delta t} \tag{7.80}$$

$$d_0 = K_I + 2K_P/\Delta t \tag{7.81}$$

$$d_1 = K_I - 2K_P/\Delta t \tag{7.82}$$

To distinguish from other dynamic elements, new symbols are introduced:

$$c_{PI}\Delta\omega = d_{PI}v_q + h_{PI} \tag{7.83}$$

where $c_{PI} = c_0$, $d_{PI} = d_0$, $h_{PI} = h_1(t - \Delta t)$.

3. The input–output relation of the saturation element is expressed by the following three equations:

$$\omega = \Delta\omega + \omega_n \tag{7.84}$$

$$\omega = \omega_{max} \tag{7.85}$$

$$\omega = \omega_{min} \tag{7.86}$$

4. The voltage-controlled oscillator is an integrator:

$$p\theta = \omega \tag{7.87}$$

Using the trapezoidal discretizing rule, we have

$$\theta(t) = \frac{\Delta t}{2}\omega(t) + \left[\theta(t - \Delta t) + \frac{\Delta t}{2}\omega(t - \Delta t)\right] \quad (7.88)$$

For brevity,

$$\theta = d_{INT}\omega + h_{INT} \quad (7.89)$$

where $d_{INT} = \Delta t/2$, $h_{INT} = \theta(t - \Delta t) + \Delta t\omega(t - \Delta t)/2$.

5. Summarizing the input–output equations for each element, the fundamental equation of the control system in the form of $f(N^{(k)}) + J^{(k)}\Delta x^{(k)} = 0$ is obtained, where

$$x = \left[v_\alpha, v_\beta, v_q, \Delta\omega, \omega, \theta\right]^T \quad (7.90)$$

$$f(x) = \begin{bmatrix} x_1 - \left(\frac{2}{3}v_a - \frac{1}{3}v_b - \frac{1}{3}v_c\right) \\ x_2 - \left(\frac{1}{\sqrt{3}}v_b - \frac{1}{\sqrt{3}}v_c\right) \\ x_3 - \left(x_1 \cos x_6 - x_2 \sin x_6\right) \\ c_{PI}x_4 - d_{PI}x_3 - h_{PI} \\ x_5 - x_4 - \omega_n \\ x_6 - d_{INT}x_5 - h_{INT} \end{bmatrix} \quad (7.91)$$

$$J^{(k)} = \begin{bmatrix} 1 & & & & & \\ & 1 & & & & \\ -\cos x_6^{(k)} & -\sin x_6^{(k)} & 1 & & & x_1^{(k)}\sin x_6^{(k)} + x_2^{(k)}\cos x_6^{(k)} \\ & & -d_{PI} & c_{PI} & & \\ & & & -1 & 1 & \\ & & & & -d_{INT} & 1 \end{bmatrix} \quad (7.92)$$

With the expression of the non-linear function $f(x)$ as (7.91) and its Jacobian matrix as (7.92), the PLL system can then be solved iteratively.

7.5 Interpolation for Power Electronics Simulation

The pervasive power electronic devices pose a major challenge to the transient simulation of distribution networks. This section briefly reviews the issues caused by these switches and the interpolation solution.

In fixed-step transient simulations, the switch status can only be changed at the end of the integration steps. This introduces a time delay in the switching instant and causes unrealistic spikes in the voltage and current wave forms, i.e. uncharacteristic harmonics. To improve the simulation accuracy, the exact switching instant must be accounted for. The simplest approach is to use small step sizes and integrate to the exact time instant of the switching. However, as we have known from the discretization process of the trapezoidal rule in Section 7.3, the expression of the equivalent conductance G_{eq} of each element already contains a step size Δt. The adaption of the step size requires updating all these conductances and regenerating the conductance matrix, which is time consuming and makes step adaption in the nodal analysis approach inefficient.

A more efficient and widely used approach is interpolation. Let us take the turnoff of a diode shown in Figure 7.11 as an example. When a current zero-crossing is detected in the simulation step of $[t - \Delta t, t]$ (as $I(t - \Delta t)$ and $I(t)$ have reversed signs), the switching instant t_D is first located using linear interpolation. Let $t_D = t - \Delta t + \alpha \Delta t$; system variables can be restored to the instant before switching using Equation 7.93. The simulation can then be continued from t_D with updated states.

$$x(t_D) = x(t - \Delta t) + \alpha(x(t) - x(t - \Delta t)) \tag{7.93}$$

The interpolation approach can restore the system variables to the switching instant without reintegration. The algorithm is simple, fast and

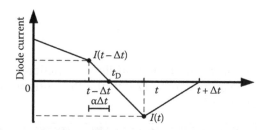

FIGURE 7.11 Linear interpolation of a diode.

effective. The assumption is that the dynamics within a step size can be linearly approximated, which is suitable when the step size is small.

In practical simulations, multiple switching could occur within the same simulation step. Generally, the switching frequency of power electronic devices is comparable to the simulation step size. As a consequence, the occurrence of multiple switching tends to increase as the scale of the simulation system becomes larger. Interested readers are referred to [7] for more in-depth information.

7.6 Case Studies

Two case studies are discussed here to show how the theory described could be used to carry out transient as well as steady state studies of distribution systems with various components.

7.6.1 Four-Node Test System

A four-node test system, as shown in Figure 7.12, is used for this case study. The parameters of the network are given in Example 6.7.

In steady-state analysis, such as power flow and short circuit calculations, only the active and reactive power of the DG system injected to Busbar B2 is used in the calculation. However, for the electromagnetic transient simulation, the detailed circuit topology and the control system need to be modelled. As shown in Figure 7.12, the PV system at Busbar B2 is modelled as eight parallel inverters, each with a rating of 625 kVA, and is connected to Busbar B2 through a transformer. The PV array is modelled as an ideal voltage source, and the inverter circuit is controlled with a dual-loop structure that has a power outer loop and a current inner loop, as shown in Figure 7.13.

The PV system output prior to the fault is 4 MW with a unit power factor. A short-circuit fault occurs at 0.3 s at Busbar B3 with each phase having grounding resistance of 5 Ω. The fault is eliminated after 5 cycles. Different types of fault are considered in the following.

7.6.1.1 Three-Phase to Ground Short-Circuit Fault

Under the three-phase to ground short-circuit fault, the voltage, current and active/reactive power wave forms at Busbar B2 are obtained and shown in Figures 7.14 through 7.16.

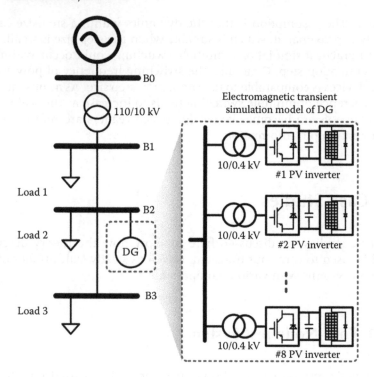

FIGURE 7.12 Diagram of a four-node test system for EMT simulation.

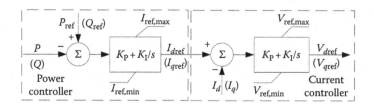

FIGURE 7.13 Dual-loop control of an inverter.

At the instant of fault occurrence, there are considerable spikes and fluctuations in the current wave form of the PV system, as shown in the dashed area in Figure 7.15a and the zoom-in plot in Figure 7.15b. Power electronic devices typically have weaker over-current capability than conventional power sources. Under severe fault conditions, this current peak could trigger the protection of the PV system and lead to an immediate shutdown of the system. In this case study, the peak current is relatively small, so the PV system keeps operating. Under the fast regulation of the

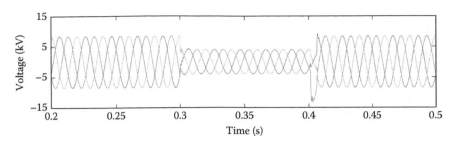

FIGURE 7.14 Voltage of Busbar B2 under a three-phase to ground fault.

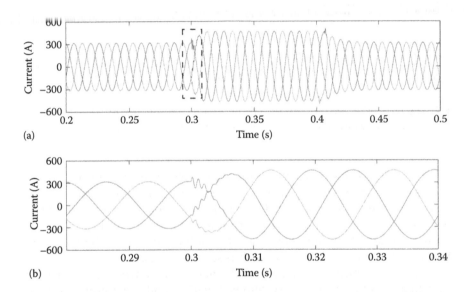

FIGURE 7.15 Current injected to Busbar B2 under a three-phase to ground fault: (a) prior, during and after the fault and (b) zoom-in at fault occurrence.

current control loop, the output current of the PV system is restored in several cycles.

The regulation effect of the power loop plays a dominant role in the ensuing transient process. Due to the reduction of the voltage level after the fault, the active power output of the PV system is significantly reduced. In compliance with the control diagram in Figure 7.13, the power loop will raise the current reference value in response so that the output current of the PV system is increased and the active power ramps up towards the set target (Figure 7.16a). This process will continue until the current reference hits the upper limit, which is set as 1.5 times the rated current value

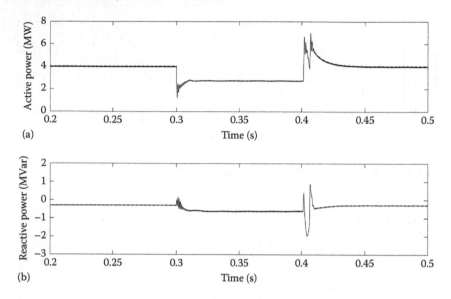

FIGURE 7.16 PQ wave form under a three-phase to ground fault: (a) active power injected to Busbar B2 and (b) reactive power injected to Busbar B2.

in this study. At this time, the system has reached the steady state under the fault condition, and the active power output of the PV system is still below the pre-fault condition. For a fault condition with a small voltage dip, the active power under the fault condition may reach the level of pre-fault condition. After fault clearance, the output current of the PV system is restored to the pre-fault level within several cycles.

It can be noted in Figure 7.16b that the reactive power injection into Busbar B2 from the PV system is negative, and the amount is larger during the faulty period. This is the reactive power consumption on the step-up transformer, and since the current magnitude is higher during the fault, the consumption is also higher.

Now let us compare the result of the electromagnetic transient simulation with the short circuit calculations carried out in Chapter 6. Figure 7.17 shows the RMS value of the voltage at Busbar B2 and the current injection to that Busbar. In Section 6.7.3, the system voltage and current under a three-phase to ground fault are computed. These values are comparable with the steady values under fault conditions shown in Figure 7.17. Table 7.2 shows the comparison of values calculated using manual calculations and simulations. Agreement between the two values is observed.

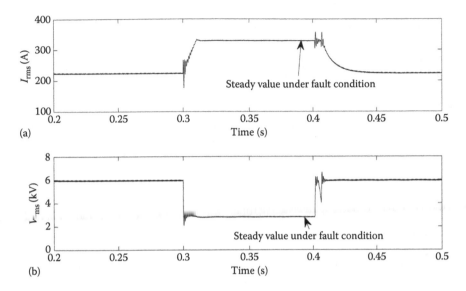

FIGURE 7.17 RMS value of Busbar B2 voltage and current injection: (a) current injection of the PV system and (b) Busbar B2 voltage.

TABLE 7.2 Simulation Result Comparison of EMT Simulation and Short-Circuit Calculation

	EMT Simulation	Short Circuit Calculation
PV injection current (A)	329.5	329.9
Busbar B2 voltage (kV)	2.821	2.818

7.6.1.2 Two-Phase Short-Circuit Fault

In recent years, a lot of research has been done on the control strategy of DG systems under unbalanced fault conditions. The transient behaviour of a DG system changes significantly under different control strategies. The specifics of these control techniques are beyond the scope of this chapter. This section only shows the fault characteristics when no additional compensation control techniques are used. The voltage, current and active/reactive power wave forms under a two-phase short-circuit fault condition at Busbar B2 are shown in Figure 7.18.

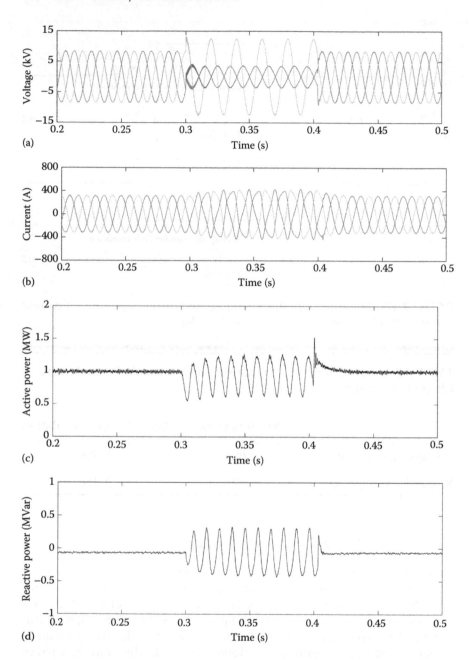

FIGURE 7.18 Simulation curves for a two-phase short circuit fault: (a) voltage of Busbar B2, (b) current injection into Busbar B2 from PV system, (c) active power injection into Busbar B2 from the PV system and (d) reactive power injection into Busbar B2 from the PV system.

As is clear from Figure 7.18a, Busbar B2 voltage is unbalanced under the two-phase fault, with the faulty phase voltage decreased and the unfaulty phase voltage increased. The output current of the PV system is distorted, with apparent negative sequence component and low order harmonics, as shown in Figure 7.18b. The power output of the PV system contains a significant 100 Hz component, as shown in Figure 7.18c and d.

7.6.2 IEEE 123-Node Test Feeder with PV Integration

The IEEE 123-node test feeder is a radial distribution network that is complex and highly unbalanced. The diagram of the system is shown in Figure 7.19. There are single-phase and three-phase lines and loads in the system. Nine PV systems with single-stage topology are dispersedly integrated into the distribution system, each with a capacity of 20 kWp. The network parameters of the 123 nodes distribution network can be found in [8], and the model and parameters of the PV systems are given in Sections 7.A.2 and 7.A.3.

The solar radiation level at each PV system is assumed to be identical. The initial output power of the PV systems is 0.5 pu, and a step change

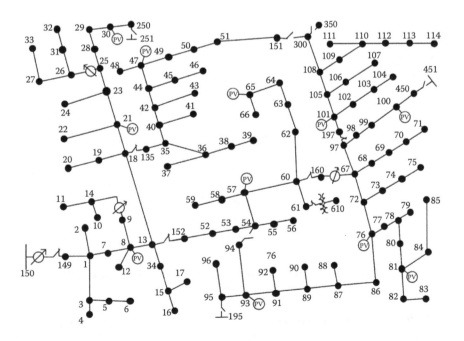

FIGURE 7.19 IEEE 123-node test feeder system with PV integration.

of the solar radiation level happens at 0.7 s, after which the PV systems have rated output power. At 1.2 s, a single-phase to ground fault occurs at Busbar 101, which is cleared after 0.3 s. Choosing the high voltage side of the step-up transformer at busbar 76 as the measurement point, the bus voltage and current injection from the PV system are shown in Figures 7.20 through 7.22.

Figure 7.20 shows the voltage and current wave form during the change in the solar radiation level at 0.7 s. As the PV systems in this case study have a relatively small capacity, the change in the output power of the PV systems has no evident influence on the bus voltage, as shown in Figure 7.20a. It is also observed from this figure that the steady-state bus voltage is slightly unbalanced, which is caused by the intrinsic unbalance in the structure and parameters of the distribution networks. This is also the reason the steady-state analysis method introduced in Chapter 6 adopts a three-phase modelling approach, which is different from conventional transmission systems. As shown in Figure 7.20b, the output current of the PV system ramps up quickly as the solar radiation steps up.

Figure 7.21 shows the voltage and current wave form during the short-circuit fault at 1.2 s. The faulty phase voltage severely drops while the other two phases see a slight increase in their voltage. On the other hand, the output current of the PV system increases significantly,

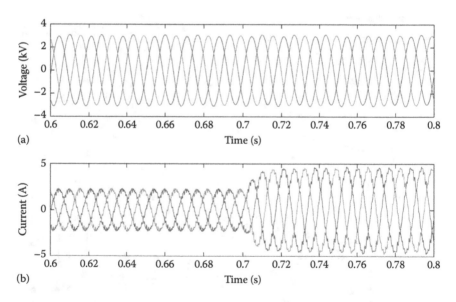

(a)

(b)

FIGURE 7.20 PV system voltage and current wave form during change in the radiation level: (a) voltage wave form at Busbar 76 and (b) current wave form of PV system.

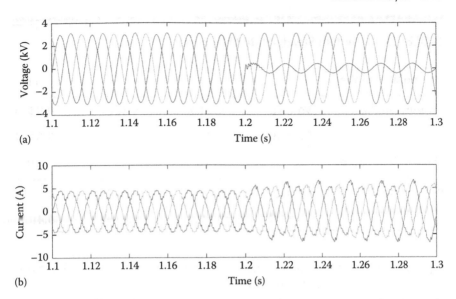

(a)

(b)

FIGURE 7.21 PV system voltage and current wave form during a single-phase fault: (a) voltage wave form at Busbar 76 and (b) current wave form of PV system.

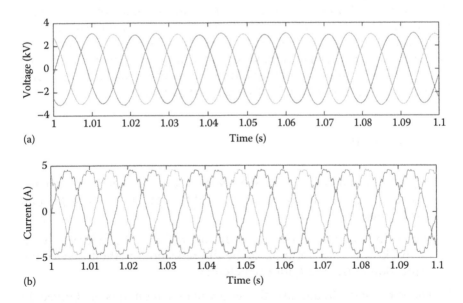

(a)

(b)

FIGURE 7.22 The PV system voltage and current wave form under steady state: (a) voltage wave form at Busbar 76 and (b) current wave form of the PV system.

TABLE 7.3 PV System Steady-State THD Level

Harmonic Number		5	7	11
Proportion (%)	Voltage	0.210	0.153	0.0493
	Current	0.256	0.138	0.0474
Harmonic Number		13	17	19
Proportion (%)	Voltage	0.0564	0.0249	0.0418
	Current	0.0629	0.0794	0.487

especially for the faulty phase, and a small portion of low-order harmonics appear in the current wave form.

In all of these simulation curves, high-frequency harmonic distortion can be observed, which is caused by the high-frequency switching power electronics with PWM. A major application area of the electromagnetic transient simulation is simulation studies concerning harmonics. The results of electromagnetic transient simulations can be used as input data for harmonic analysis programs for further quantitative analysis. Figure 7.22 shows a time window of the steady-state voltage and current wave forms at Busbar 76. A frequency analysis is conducted on this window and the results are listed in Table 7.3, where the total harmonic distortion (THD) is used as the index to measure the harmonic level. The THD is defined as the ratio of the sum of the powers of all harmonic components to the power of the fundamental frequency.

☐ Questions

1. Explain the major differences between the nodal analysis approach and the state space analysis approach. What are their advantages and disadvantages?

2. Derive the equivalent circuit model for a linear RLC element with the backward Euler formula as the discretization method. Derive the expression of G_{eq} and I_h for each element.

3. For the boost circuit, use the equivalent circuit model derived in Example 7.1 to establish the MANA equations, assuming the power electronic switches are ideal.

Appendix 7A

7.A.1 First Example

1. Boost circuit parameters and simulation parameters (Table 7.4)

2. MATLAB Code Excerpt

```
% define parameters (omitted, see table in section 7.7.1)
% initial value setting
x = zeros(2,round(Tend/dt)+1); % zero initial value
starting

%% simulation stepping
for n = 1:round(Tend/dt)
    t = n*dt;
    if mod(n,round(T/dt)) < round(D*T/dt)
        A = A1; B = B1; % circuit works in mode 1
    else
        A = A2; B = B2; % circuit works in mode 2
    end
    x(:,n+1) = (eye(2)-dt/2*A)\...
        (eye(2)+dt/2*A)*x(:,n)+dt*B*vs); % trapezoidal
        method
end

%% display simulation results
figure(1)
plot(1e3*[0:dt:Tend],x(1,:))
xlabel('time (ms)')
ylabel('current (A)')
title('Inductor Current')

%% more visualization of simulation results (omitted)
```

7.A.2 Transient Simulation Model of a PV System

PV generation systems have been previously introduced in other parts of the book. Additional modelling details are needed for the electromagnetic transient simulation. The diagram of a typical PV system is shown in Figure 7.23.

As in Figure 7.23, the electrical part of the PV system includes PV arrays, a DC capacitor, an inverter circuit and LC filters. The equivalent

TABLE 7.4 Boost Circuit Parameters and Simulation Parameters

Circuit parameter		Control parameter	
V_s	100 (V)	Control period T	0.2 (ms)
Inductor L	10 (mH)	Duty cycle D	30%
Capacitor C	500 (µF)	Simulation parameter	
		Step size Δt	10 (µs)
Load resistor R	10 (Ω)	Simulation time T_{end}	50 (ms)

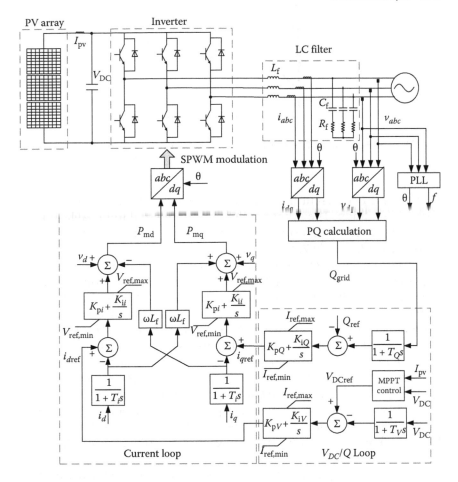

FIGURE 7.23 Diagram of a single-stage, grid-connected PV generation system.

circuit model of the PV array is introduced in Chapter 2. For the sake of completeness, the equivalent circuit model is shown in Figure 7.24. The photocurrent I_{ph} in the figure satisfies

$$I_{ph} = I_{sc,ref}\left[\frac{S}{S_{ref}} + \frac{J}{100}\left(T - T_{ref}\right)\right];\qquad(7.94)$$

The diode current I_d satisfies

$$I_d = I_{sref}\left(\frac{T}{T_{ref}}\right)^3 e^{\frac{qE_g}{Ak}\left(\frac{1}{T_{ref}} - \frac{1}{T}\right)}\left(e^{\frac{q(V+IR_s)}{AkT}} - 1\right)\qquad(7.95)$$

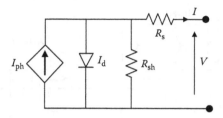

FIGURE 7.24 Equivalent circuit model of the PV array.

where S and T represent the solar radiation level and temperature of the working condition. The other parameters are referred to in Table 7.5.

The modelling and simulation of the three-phase inverter circuit is referred to in Example 7.2.

The control system adopts a dual-loop control structure. The outer loop implements the MPPT control and the reactive power control. Compared with the dual-loop structure in Figure 7.13, the MPPT control is unique to the PV system. The reference signal of the DC voltage V_{DCref} is computed based on the current I_{pv} and the voltage V_{DC} of the PV array. The measured DC voltage V_{DC} and the reactive power of the inverter are filtered and compared to their reference signals V_{DCref} and Q_{ref}. The deviations are corrected by the PI block, which generates the reference signals i_{dref} and i_{qref} for the inner current loop. When the working condition is changed (solar radiation, ambient temperature or the network topology), the deviation signal of the DC voltage or reactive power is non-zero, and the PI block will adjust the current reference accordingly until the deviation is eliminated.

The inner loop implements the current control under the dq rotation framework, which transforms the three-phase instantaneous signal into DC signals through Park transformation. In Figure 7.23, the three-phase instantaneous current i_{abc} is transformed to the dq-axis current i_{dq}, which is filtered and then compared with the reference signals i_{dref} and i_{qref}. The deviations in the current signal are regulated by the PI block, which also includes saturation values to limit the maximum output current of the inverter. The output of the PI block is compensated by the voltage feed-through term and cross-coupling term to obtain the voltage modulation signals P_{md} and P_{mq}. In this control structure, the purpose of the voltage feed-through and the cross-coupling compensation is to decouple the dq component to realize the separation of the control objectives. The modulation signals P_{md} and P_{mq} are then transformed back to the three-phase quantities to fire individual phases of

TABLE 7.5 PV Array Parameters

Parameter	Description	Value
S_{ref}	Solar radiation level at the standard condition	1000 W/m²
T_{ref}	Working temperature at the standard condition	298 K
$V_{OC,ref}$	Open-circuit voltage at the standard condition	21.7 V
$I_{SC,ref}$	Short-circuit current at the standard condition	3.35 A
$V_{mp,ref}$	Maximum power point voltage at the standard condition	17.4 V
$I_{mp,ref}$	Maximum power point current at the standard condition	3.05 A
$P_{mp,ref}$	Maximum power at the standard condition	53.07 W
J	Temperature coefficient of the photocurrent	0.065%
R_s	Series resistance of the PV cell	0.312 Ω
R_{sh}	Parallel resistance of the PV cell	∞ (none)
k	Boltzmann's constant	1.38e−23 J/K
q	Coulomb's constant	1.60e−19 C
E_g	Band-energy gap	1.237 eV
A	Fitted coefficient of the diode characteristics	1.5
N_s	Number of modulus in series	20
N_p	Number of modulus in parallel	9

the inverter circuit. The current loop converts the current reference signal to the voltage reference signal and is the fast response part of the inverter controller.

Another important part of the control system is the phase-locked loop, which is introduced in Example 7.4.

7.A.3 Model Parameters of the 20 kW PV System

1. PV array parameters

2. Electrical circuit parameters of the PV system (Table 7.6)

3. Control system parameters of the PV system (Table 7.7)

TABLE 7.6 Electrical Circuit Parameters of the PV System

Parameter	Description	Value
C	DC capacitor	2500 μF
LC filter		
L_f	Filter inductor	4 mH
C_f	Filter capacitor	50 μF
R_f	Series resistance of the filter capacitor branch	1.0 Ω
Step-up transformer		
S_T	Rating of the transformer	50 kVA
V_1/V_2	Winding voltages	4.0/0.18 kV
v_k	Short-circuit voltage	8%
r_k	Copper losses	3.1%

TABLE 7.7 Control System Parameters of the PV System

Parameter	Description	Value
Outer control loop		
T_V, T_Q	Filter time constants	0 (no filter)
K_{pV}	Voltage loop K_p	1
K_{iV}	Voltage loop K_i	2
K_{pQ}	Reactive power loop K_p	0.01
K_{iQ}	Reactive power loop K_i	0.1
$I_{ref,max}/I_{ref,max}$	Saturation of current reference	±1.5 (pu)
Inner control loop		
T_i	Filter time constants	0 (no filter)
K_{pi}	Current loop K_p	2
K_{ii}	Current loop K_i	5
$V_{ref,max}/V_{ref,max}$	Saturation of voltage reference	±1.5 (pu)
Phase-locked loop (see Figure 7.10)		
K_P	Loop filter k_p	1.8
K_I	Loop filter k_i	32

☐ References

1. C. W. Ho, A. E. Ruehli and P. A. Brennan, The modified nodal approach to network analysis, *IEEE Transactions on Circuits and Systems*, 22(6), 504–509, 1975.
2. J. Mahseredjian, S. Dennetière, L. Dubé et al., On a new approach for the simulation of transients in power systems, *Electric Power Systems Research*, 77(11), 1514–1520, 2007.

3. H. W. Dommel, *Electromagnetic Transients Program: Reference Manual: (EMTP Theory Book)*, 2nd edn., Vancouver, BC, Canada: MicroTran Power System Analysis Corp., 1996.
4. N. Watson and J. Arrillaga, *Power Systems Electromagnetic Transients Simulation*. London, UK: The Institution of Electrical Engineers, 2003.
5. K. S. Kundert, Sparse 1.3 – A sparse linear equation solver, 1986. Available from: http://www.netlib.org/sparse/, accessed on September 11, 2016.
6. T. A. Davis and E. P. Natarajan, Algorithm 907: KLU, a direct sparse solver for circuit simulation problems, *ACM Transaction on Mathematical Software*, 37(3), 1–36, 2010. Available from: http://www.cise.ufl.edu/research/sparse/klu/, accessed on September 11, 2016.
7. M. Zou, J. Mahseredjian, G. Joos et al., Interpolation and reinitialization in time-domain simulation of power electronics circuits, *Electric Power Systems Research*, 76(8), 688–694, 2006.
8. W. H. Kersting, Radial distribution test feeders, *IEEE Transactions on Power Systems*, 6(3), 975–985, 1991.

Control and Protection of Microgrids

8.1 Introduction

A microgrid can contain different types of distributed energy resources (DERs), such as distributed generators (which are also called micro-sources in a microgrid) and energy storage systems. Due to the intermittent nature of renewable energy sources, managing a microgrid is different from managing conventional power grid. It has different source composition and structure and operation mode, and can be operated in either grid-connected or islanded mode. Therefore, a microgrid has its own control and protection characteristics.

8.2 Control of Microgrids

8.2.1 Control Philosophy

As a microgrid can normally operate either in grid-connected or islanded mode, a reliable, seamless transition between these two modes should be achieved.

In the grid-connected mode, as the voltage and frequency in a microgrid are dominated by the grid, the constant power control (*PQ* control) or a droop control strategy can be used for distributed generators (DGs), which is similar to the scenario that DGs are connected to a distribution system directly.

In the islanded mode, maintaining the microgrid stability needs the coordinated control of the DGs and energy storage systems. Generally, there are two main control strategies to control the voltage and frequency of a microgrid, named as master-slave and peer-to-peer control.

8.2.1.1 Master-Slave Control

In an islanded microgrid that adopts master-slave control, one distributed generation (or energy storage system) adopts constant voltage and frequency control (*V/f* control) [1] that provide a reference voltage and frequency to other DGs, while the other DGs adopt *PQ* control [2]. As shown in Figure 8.1, the controller which adopts *V/f* control is called the master controller, while the others are called the slave controllers.

It should be note that when a master-slave controlled microgrid operating in the grid-connected mode, all DGs should adopt *PQ* control or droop control. Once the microgrid is transferred to the islanded mode, the master controller should immediately be switched to *V/f* control. In this situation, the variation of loads is mainly compensated by the master DG, and thus the master unit is supposed to have enough power capacity and fast load response capability.

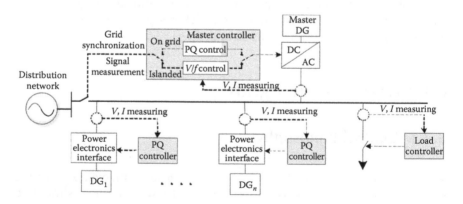

FIGURE 8.1 A microgrid using the master-slave control strategy.

8.2.1.2 Peer-to-Peer Control

Peer-to-peer control means all DG units operate in a similar control scheme to control the voltage and frequency simultaneously, which only uses the local measurements [3]. As shown in Figure 8.2, an appropriate control for each DG is the droop control, which can share load variation and realize the rational distribution of power [4–6] automatically according to the droop coefficient.

Compared with the master-slave control, the peer-to-peer control introduces local control of the output of DGs, which is beneficial for DGs' plug-and-play and is also convenient for the integration of different types of DGs. Furthermore, a seamless switching of modes, i.e. between grid-connected and islanded, can be realized for a microgrid using a peer-to-peer control, as the DG controller can keep using droop control unchanged during the transition process.

8.2.2 Control Schemes for Inverter-Based DGs

The stable operation of a microgrid relies on the effective control of DGs. DGs can be classified into three categories: inverter-based, synchronous and asynchronous. The control and synchronization technologies of small-scale synchronous generators are relatively mature and the control of asynchronous generators is relatively simple. Most of the DGs in a microgrid, such as photovoltaic systems, fuel cells and micro gas turbines, are normally require power conversion devices as the grid-connection interface, which can be called are inverter-based DGs.

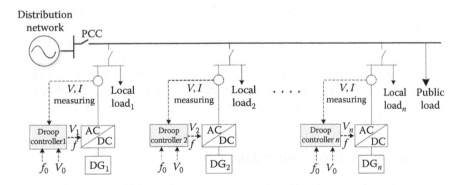

FIGURE 8.2 A microgrid using a peer-to-peer control strategy.

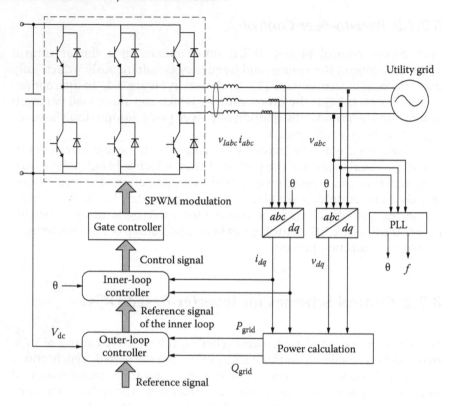

FIGURE 8.3 Typical structure of a three-phase inverter control system.

There are a variety of control methods for inverters, among which dual-loop control is widely used, where the outer-loop controller is used to track different control purposes, with its output as a reference signal of the inner loop. The dynamic response of the outer loop is generally slow. On the contrary, the inner-loop controller is usually designed to have a fast dynamic response to achieve better tracking of the output current to the reference signal. For a lower requirement of dynamic response performance, a simpler design with merely the outer-loop controller can be devised. A typical structure of a three-phase inverter control system is shown in Figure 8.3.

8.2.2.1 Outer-Loop Controller

Depending on the roles DGs play in a microgrid, different controls are adopted by different inverters, including constant power control (i.e. *PQ* control), constant voltage and frequency control (i.e. *V/f* control) and droop control. Each of these controls has a different control objective of the outer-loop controller.

8.2.2.1.1 PQ Control

The objective of *PQ* control is to control the outputs of the active and reactive power of DGs and ensure they are equal to the reference *P/Q* values. The control principle is shown in Figure 8.4.

Suppose the initial operating point of a DG is at A, where the active and reactive power are equal to the references P_{ref} and Q_{ref}, respectively. At point A, the frequency is f_0 and the voltage at the connection point of the DG is v_0. When the frequency varies in a range ($f_{min} \leq f \leq f_{max}$), the active power controller adjusts the frequency characteristic curve to maintain the output active power of the DG at the given reference P_{ref}. When the voltage changes in a range ($v_{min} \leq v \leq v_{max}$), the reactive power controller also adjusts the voltage characteristic curve to maintain the output reactive power of DG at Q_{ref}. Therefore, DGs with *PQ* control cannot maintain the frequency and voltage of the network, but only ensure that their *P/Q* outputs meet the corresponding requirements. Figure 8.5 shows a typical structure of the outer-loop controller with *PQ* control.

As shown in Figure 8.5, three-phase instantaneous currents i_{abc} and voltages v_{abc} are first translated to i_{dq} and v_{dq} in the *dq* reference frame through Park transformation. The instantaneous active and reactive output power P_{grid} and Q_{grid} are calculated and processed by a low-pass filter, in order to obtain the average active and reactive power, P_{filt} and Q_{filt}. These average values are then compared with the reference signals P_{ref} and Q_{ref}. The errors are sent to a PI controller where reference signals i_{dref} and i_{qref} of the inner-loop controller are produced. The PI controller keeps integrating

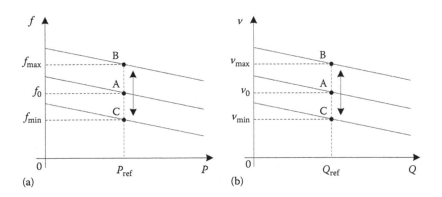

FIGURE 8.4 Principle of *PQ* control: (a) active power and frequency and (b) reactive power and voltage.

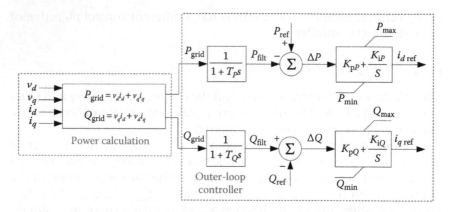

FIGURE 8.5 Typical structure of an outer-loop controller with *PQ* control.

until the error signal becomes zero, and this will ensure the output *P/Q* values equal to the reference signals P_{ref} and Q_{ref}. Theoretically, the controller can reach a steady state indicating that the output power is equal to the reference, which corresponds to the situation that the operation point moves in vertical direction shown in Figure 8.4.

It is worth pointing out that the output power of DGs on the *dq* frame can be calculated as

$$\begin{cases} P_{grid} = 1.5(v_d i_d + v_q i_q) \\ Q_{grid} = 1.5(v_q i_d - v_d i_q) \end{cases} \tag{8.1}$$

In the rotating reference frame (*dq* frame), as shown in Figure 8.6, the voltage on the *q*-axis is equal to zero when selecting the *d*-axis orientated toward the voltage vector. Under this condition, the active power is only

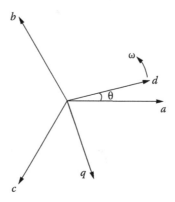

FIGURE 8.6 *abc* reference frame and *dq0* reference frame.

related to the active current on the d-axis, while the reactive power is merely related to the reactive current on the q-axis. Thus the current references can be calculated by the power references and the voltage of the AC network, as shown in (8.2), which means that only an inner-loop controller is needed to realize the PQ control purpose:

$$\begin{cases} P_{\text{grid}} = 1.5 v_d i_d \\ Q_{\text{grid}} = -1.5 v_d i_q \end{cases} \Rightarrow \begin{cases} i_{d\,\text{ref}} = \dfrac{P_{\text{ref}}}{1.5 v_d} \\ i_{q\,\text{ref}} = -\dfrac{Q_{\text{ref}}}{1.5 v_d} \end{cases} \tag{8.2}$$

Due to the decoupled control of the active and reactive power, other control strategies can be derived if one or two inputs and reference signals are changed, such as constant DC voltage, reactive power control [7–10], inverter output voltage control [11,12] and simplified constant power control [13].

8.2.2.1.2 V/f Control

The purpose of V/f control is to maintain the voltage magnitude and frequency at the connection point of the DG. In general, a DG with V/f control, which is similar to the slack bus in a conventional power system, can provide voltage and frequency support for a microgrid operating in an islanded mode. The power generated from each DG is limited by its capacity, so the power balancing between the demand and generation in islanded operation should be guaranteed in advance when using this control.

The principle of V/f control is illustrated in Figure 8.7. Assuming that an initial operating point of a DG is at A, the frequency and voltage at the AC bus are f_{ref} and V_{ref}, and the output of the active power and reactive power of the DG are P_0 and Q_0. The objective of V/f control is to maintain voltage and frequency constant.

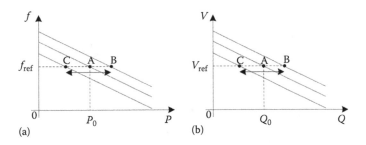

FIGURE 8.7 V/f control principle: (a) active power and frequency and (b) reactive power and voltage.

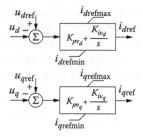

FIGURE 8.8 Outer-loop control structure for voltage control.

When the PWM modulation method is used in DC/AC converters, the frequency of the system is decided by the modulated signal frequency, instead of using a frequency feedback controller. Therefore, only a feedback controller is needed for the voltage control in *V/f* control mode. A typical outer-loop control structure for the voltage control is shown in Figure 8.8.

The output frequency of the inverter is generated through a reference sinusoidal oscillator, which is considered as the directional reference vector for the oriented direction of the *d*-axis of the synchronous rotating frame. v_{dref} and v_{qref} represent the *d*-axis and *q*-axis components reference of AC bus voltage in the microgrid. v_{dref} is equal to 311 V and v_{qref} is equal to 0 V with a rated phase voltage of 220 V.

8.2.2.1.3 Droop Control

Droop control is a method that imitates the static power frequency characteristic of generators [14,15], and the control principle is shown in Figure 8.9.

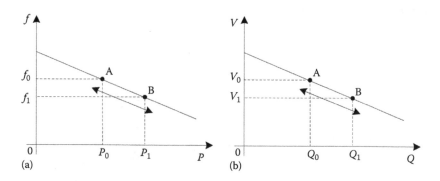

FIGURE 8.9 Droop control principle: (a) active power and frequency and (b) reactive power and voltage.

Assuming that an initial operating point of a DG is at A, the output of the active and reactive power are P_0 and Q_0, and the frequency and voltage at the DG connection point are f_0 and V_0. The frequency decreases due to the lack of active power and the voltage magnitude decreases due to the lack of reactive power, and vice versa. For example, if the frequency decreases along with the increase of active load, the control system increases the output power of the DG according to the droop characteristic. Consequently, this leads to a new power balance as shown at point B in Figure 8.9a. The droop curves of active power-frequency and reactive power-voltage are expressed by

$$\begin{cases} P = P_0 + \left(f_0 - f \right) K_f, \\ Q = Q_0 + \left(V_0 - V \right) K_V. \end{cases} \tag{8.3}$$

or

$$\begin{cases} f = f_0 + \left(P_0 - P \right) K_P, \\ V = V_0 + \left(Q_0 - Q \right) K_Q. \end{cases} \tag{8.4}$$

There are two basic droop control methods: (1) controlling the output power via the frequency and magnitude of the voltage, which is known as $(f$-$P)/(V$-$Q)$ droop control [16], and (2) controlling the frequency and magnitude of the voltage based on the output power, which is known as $(P$-$f)/(Q$-$V)$ droop control. Here the latter is taken for a detailed analysis, and the typical structure of an outer-loop controller is shown in Figure 8.10.

As shown in Figure 8.10, with a single control loop, the frequency and voltage magnitude are managed by $(P$-$f)/(Q$-$V)$ droop control. The

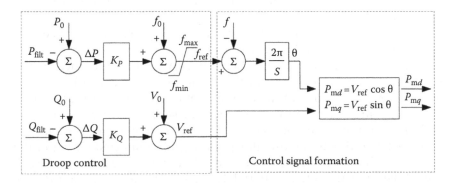

FIGURE 8.10 Typical structure of an outer-loop controller based on $(P$-$f)/(Q$-$V)$ droop control.

output active power is used to regulate the voltage phase through the integration of the reference frequency, and the output reactive power is applied to control the output voltage magnitude. The output signals of the controller are P_{md} and P_{mq}, which can be applied to modulate the dq-axis components of the voltage. The modulation principle can refer to the relevant instructions in the inner-loop controller part. As the unbalanced load or non-linear load has an inevitable influence on the output voltage of the inverter when only a single loop control mode is adopted, the voltage control loop can be added into the droop control method to improve the quality of the output voltage [17,18].

Besides these control methods, there are other control methods for droop control. For example, a virtual impedance method which adopts the Q-L droop characteristic instead of the Q-V characteristic to achieve voltage control [19]. In Reference 20, a $(P$-$V)/(Q$-$f)$ control method was also adopted, and the resistance characteristic of the low-voltage distribution lines was taken into consideration.

8.2.2.2 Inner-Loop Controller

The main aim of using an inner-loop controller is to regulate the current injected into the network in order to track the current references, and this can improve both power quality and operating performance of the system. Depending on the coordinate system, the inner-loop controller can be divided into three categories [7,8,21–23]: (1) the most widely used $dq0$ rotation coordinate system control, (2) the $\alpha\beta0$ static coordinate system control and (3) the abc natural coordinate system control. The typical structure of an inner-loop controller in $dq0$ rotation frame is shown in Figure 8.11 [7,8]. Three-phase instantaneous signals are transformed into the values under the $dq0$ rotating coordinate system. A three-phase control problem is then converted into a two-phase control problem.

The three-phase instantaneous currents i_{abc} are transformed into dq-axis components i_{dq} by Park transformation. Then i_{dfilt} and i_{qfilt} are obtained by a low-pass filter and compared with the reference signals i_{dref} and i_{qref}. The errors are regulated through a PI control regulator within the maximum output current limit. Then the output voltage signals P'_{md} and P'_{mq} are derived via feed-forward voltage compensation and cross-coupling compensation (where ωL represents the equivalent reactance [including the filter] of the grid-connected circuit of the inverter). Finally, in order to ensure a linear modulation range with a modulation ratio between 0 and 1, the amplitude of P'_{md} and P'_{mq} are limited through a module value limiter and then the modulated signals P_{md} and P_{mq} are obtained [24].

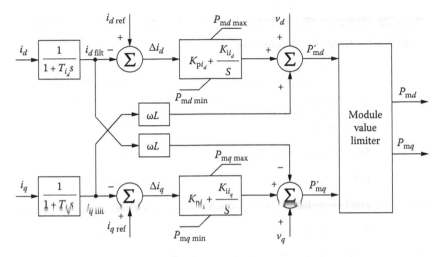

FIGURE 8.11 Typical structure of an inner-loop controller in the $dq0$ rotation frame.

8.2.3 Case Analysis

This section presents the implementation of the control strategies applied in the simple microgrid shown in Figure 8.12. This microgrid consists of two simplified distributed generators (DG_1 and DG_2), with their dynamic

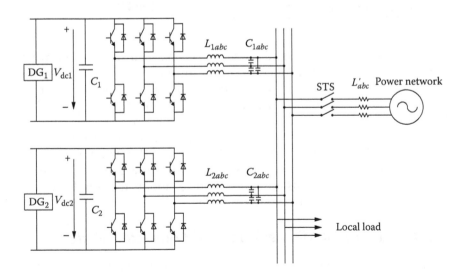

FIGURE 8.12 A microgrid system.

TABLE 8.1 Microgrid Parameters

Source	Parameter Name	Parameter Value
DG_1	Grid-connected inverter switching frequency	4 kHz
	Filter inductance (L_{1abc})	0.25 mH
	Filter capacitance (C_{1abc})	250 µF
	DC bus voltage	750 V
	Rated capacity	300 kVA
DG_2	Grid-connected inverter switching frequency	4 kHz
	Filter inductance (L_{2abc})	0.25 mH
	Filter capacitance (C_{2abc})	250 µF
	DC bus voltage	750 V
	Rated capacity	100 kVA
Power grid voltage/frequency/inductance L'_{abc}		380 V/50 Hz/30 mH
Local constant power load		300 kW

response characteristic ignored. Each DG is assumed to be an ideal DC voltage source, which is connected to the AC bus through a three-phase grid-connected inverter and an LC filter. Usually the microgrid is connected to the distribution network by a static control switch (STS). The load in the microgrid is assumed to be a constant power load, and the utility distribution network can be considered as an ideal three-phase controlled voltage source with a small reactance. The parameters of the microgrid and control systems are shown in Tables 8.1 and 8.2, respectively. In this case, the constant power control adopts the simplified control mode without the outer-loop controller.

8.2.3.1 Control of Microgrid in Grid-Connected Mode

When the microgrid is connected to the local distribution system, the main control target in the grid-connected operation mode is to

TABLE 8.2 Controller Parameters

	Outer-Loop Controller	Inner-Loop Controller
Constant power control	No outer loop controller	$K_{pi_d}=K_{pi_q}=0.15,$ $K_{ii_d}=K_{ii_q}=1.154$
Constant voltage/ frequency control	$K_{pv_d}=K_{pv_q}=2, K_{iv_d}=$ $K_{iv_q}=40$	$K_{pi_d}=K_{pi_q}=0.15,$ $K_{ii_d}=K_{ii_q}=1.154$
Droop coefficient	$K_{p1}=K_{Q1}=-0.0005,$ $K'_{p2}=K'_{Q2}=-0.001$ $P_{01}=P_{02}=0, f_{01}=$ $f_{02}=50.05$ Hz $Q_{01}=Q_{02}=0, V_{01}=$ $V_{02}=311$ V	$K_{pi_d}=K_{pi_q}=0.15,$ $K'_{ii_d}=K'_{ii_q}=1.154$

maintain the output power of the tie line between the microgrid and external network within a certain range. Hence DG_1 and DG_2 both adopt constant power control mode, and the inner-loop controller is shown in Figure 8.11, in which the d-axis and q-axis current reference values are directly calculated from Equation 8.2. The constant power control structure used in DG_1 and DG_2 is shown in Figure 8.13, which uses the single loop structure with inner-loop control and ignores the low-pass filter. In the constant power control, the grid voltage vector orientation is used. The grid voltage phase θ obtained by the PLL is regarded as the reference voltage phase of space vector pulse width modulation (SVPWM) and coordinate transformation.

In this scenario, the power control of the tie line is realized by adjusting the output power of DG_1. As shown in Figure 8.14a, before time t_1, the output power of both DGs is 100 kW, the microgrid load is 300 kW, and the microgrid absorbs 100 kW of active power from the external network (power of the tie line is negative, which indicates that the microgrid absorbs active power from the external network). At t_1 and t_2, the active power of DG_1 is increased by 50 kW. This reduces the power of the tie line. The three-phase instantaneous current waveform of microgrid tie line is shown in Figure 8.14b. During the actual grid-connected operation of the microgrid and power control of the tie line, the reference value of the output power of the controllable DG is always adjusted by the microgrid's central controller according to the corresponding energy management strategies and tie line power control objectives.

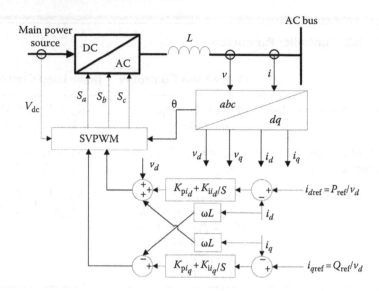

FIGURE 8.13 Constant power control structure diagram.

8.2.3.2 Control of the Microgrid in Islanded Mode

In this operation mode, two control strategies can be used. In a master-slave control, DG_1 operates as the main power supply with V/f control, establishing the voltage and frequency of the microgrid; while DG_2 operates as an auxiliary power supply with PQ control. In a peer-to-peer control, DG_1 and DG_2 both adopt $(P\text{-}f)/(Q\text{-}V)$ droop control method, maintaining the microgrid's voltage and frequency.

8.2.3.2.1 Master-Slave Control

In the master-slave control, the control structure of the slave unit DG_2 is the same as that shown in Figure 8.13, and the control structure of the master unit DG_1 is shown in Figure 8.15. In V/f control, the angular frequency of the voltage is $\omega = 2\pi * 50$ rad/s, and θ is obtained by integrating ω. This is the reference voltage phase of SVPWM and coordinate transformation.

During normal operation, the load power is 300 kW, and the reference value of DG_2 output active power is 100 kW. Thus DG_1 using V/f control outputs 200 kW. The output wave form is shown in Figure 8.16. At t_3, when the load is suddenly reduced by 150 kW. The output power of DG_2 using PQ control keeps constant, while the output of DG_1 adopting V/f control quickly reduces to 50 kW to follow the load variation, as shown in Figure 8.16a. The output current wave form of DG_1 is shown in Figure 8.16b.

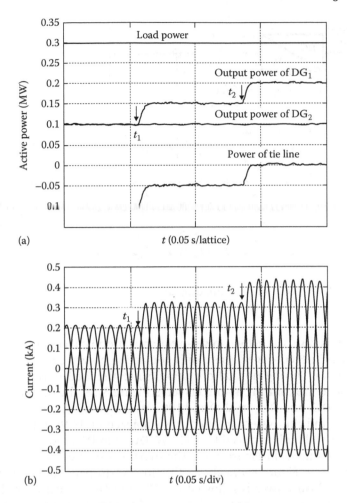

FIGURE 8.14 Power control wave form of the microgrid tie line: (a) active power of load, DGs and tie line and (b) three-phase instantaneous current waveform of microgrid tie line.

8.2.3.2.2 Peer-to-Peer Control

In peer-to-peer control, DG_1 and DG_2 both adopt $(P\text{-}f)/(Q\text{-}V)$ droop control shown in Figure 8.17, establishing the microgrid's voltage and frequency and sharing the load fluctuations. The references of voltage magnitude and frequency are obtained from the droop control loop in Figure 8.10, and the other control loops are the same as the control structures of the main power supply in the master-slave mode.

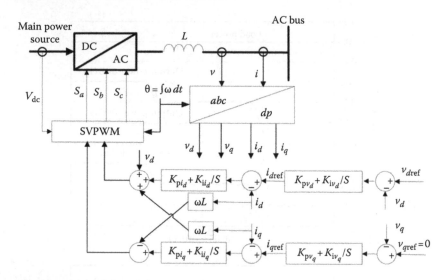

FIGURE 8.15 *V/f* control structure diagram.

In this case, the active power-frequency droop curves of DG_1 and DG_2 are shown in Figure 8.18 (here only the active power control is taken as an example). During normal operation, since the load power is 300 kW, according to the control parameters set in Figure 8.18, the output power P_{b0} and P_{a0} of DG_1 and DG_2 are 200 and 100 kW. DG_1 and DG_2 operate at points b_0 and a_0, and the system frequency is stabilized at 49.95 Hz. At t_4, when the load is suddenly reduced by 150 kW, DG_1 and DG_2 move to the points b and a, as shown in Figure 8.18. The dynamics of the system are shown in Figure 8.19. Figure 8.19a shows that in peer-to-peer control mode, the changes of load power is followed by DG_1 and DG_2 synchronously. Also, according to the droop characteristic of $P-f$, the frequency of the microgrid will increase with the decrease of the load, as shown in Figure 8.19b.

The adoption of peer-to-peer control can solve the issues of the stability control of voltage and frequency and the rational distribution of output power at the same time. However, the steady-state voltage and frequency will vary along with the change of load. Then the microgrid central controller can be used to carry out the secondary regulation of the DG droop curve to ensure that the voltage and frequency meet the requirements of reliable operation.

8.2.3.3 Switching of the Microgrid Operation Mode

During the grid-connected operation, DG_1 and DG_2 both adopt *PQ* control, and when the microgrid is switched to an islanded operation mode,

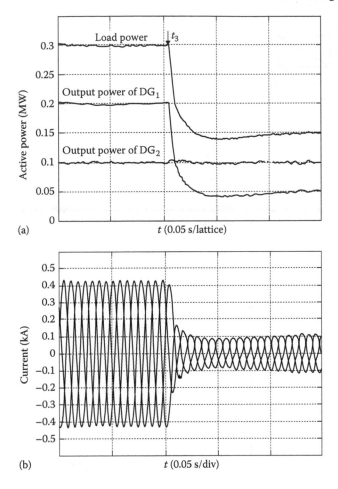

FIGURE 8.16 Microgrid master-slave control wave form: (a) active power of load and DGs and (b) output currents waveform of DG_1.

the control of DG_1 needs switching to *V/f* control (considering the master-slave control strategy is used by the DGs).

When the microgrid is switched from a grid-connected to an islanded mode, the output power of the DGs and the voltage at the AC bus will be as shown in Figure 8.20a and b. When switching the system operation mode, it is assumed that a fault detection time of 10 ms is required after an external network fault, and then the main power supply of the microgrid begins to switch the operation mode. As shown in Figure 8.20a, before t_s, the main power supply and auxiliary power supply of the grid-connected microgrid generate an active power of 150 and 100 kW, respectively; the load is 300 kW and the power of the tie

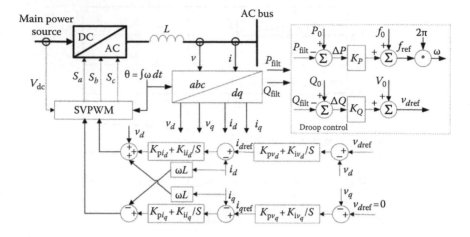

FIGURE 8.17 Droop control structure diagram.

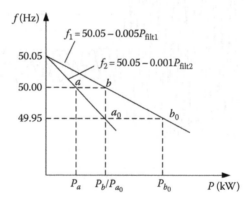

FIGURE 8.18 DG$_1$ and DG$_2$ droop control curve.

line is negative 50 kW (the microgrid needs to absorb 50 kW active power from the external network). At t_5, the STS is off due to power grid fault (voltage drop) and completely disconnected after about a half cycle (10 ms). Although voltage drop can decrease the active power of local load, in view of the fact that the local load in this case is a constant power load, the load power will be restored to 300 kW after a brief adjustment. When the main power supply of the microgrid is switched from *PQ* control to *V*/*f* control, the voltage of the AC bus is gradually recovered, as shown in Figure 8.20b. Finally, a seamless transition control of the microgrid operation mode is achieved.

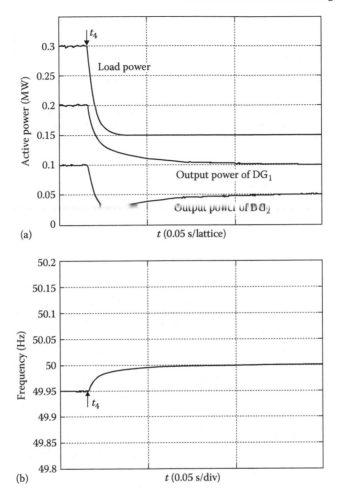

FIGURE 8.19 Wave form of microgrid with peer-to-peer control: (a) active power of load and DGs and (b) frequency of microgrid.

8.3 Protection of Microgrids

Due to the time-varying characteristics of a microgrid considering the operation modes and the network topology, the protection of microgids is facing some technical issues that deserve more attention [25]. Figure 8.21 shows a typical topology of a microgrid with several potential fault locations on the feeders.

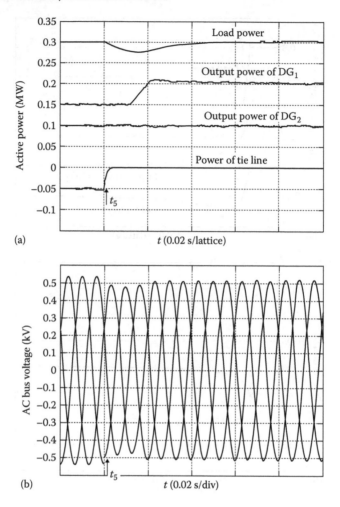

FIGURE 8.20 Wave form for microgrid switching from grid-connected to islanded mode: (a) active power of load, DGs and tie line and (b) AC bus voltage.

8.3.1 Requirements of Microgrid Protection

Three different cases are considered to describe protection issues:

1. Given that distributed generators (DGs) in the microgrid influence the conventional protection scheme, a reliable operation of the main-system-side protection should be guaranteed, and this protection scheme should also cooperate with the consumer-side protection.

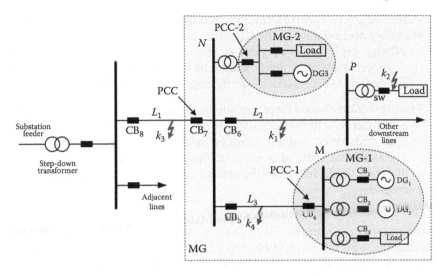

FIGURE 8.21 Typical topology of a microgrid with several potential fault locations.

For example, when a fault occurs at point k_1 in Figure 8.21, the fault current flowing through switch CB_6 on line L_2 becomes larger compared with the condition with no distributed generation, due to the infeed current from the distributed generations (DG_1 and DG_2). On the contrary, the fault current flowing through the upstream line L_1 becomes smaller because of the voltage-support effect from the distributed generation. Therefore, the conventional protection coordination of the feeders should be reconfigured.

Meanwhile, when a fault occurs on an arbitrary line at the consumer side (e.g. k_2), the reliability of the protection and breaking device SW should be checked, because the fault current becomes larger with distributed generation connected to the distribution system.

2. The protection strategy of microgrids should be able to respond to the fault in both the main power grid and the microgrid in time. When a fault in the main power grid is detected, the microgrid ought to be isolated from the main power grid. When a fault in a microgrid is detected, the faulty section should be removed to ensure power supply to the main power grid and the microgrid.

For example, the downstream system of feeder L_1 would become a microgrid in islanded mode when the circuit breakers CB_7 and CB_8 at the two terminals of feeder L_1 are tripped due to the occurrence of fault at point k_3 as shown in Figure 8.21. The islanding protection of the microgrid should detect this fault quickly and the control system should adjust its control strategies in order to make the microgrid operate safely and reliably in the islanded mode.

When a fault occurs at point k_4, the circuit breakers CB_4 and CB_5 at the two terminals of feeder L_5 should be tripped. Thus the microgrid can be divided into several small-size microgrids, such as MG-1 and MG-2.

3. The protection should have the ability to adapt to the operation modes of a microgrid, which means the protection strategy must be effective in both grid-connected and islanded modes.

    The fault occurring at point k_4 in Figure 8.21 is taken as an example to illustrate that a self-adaptive capability is needed in the protection of microgrids.

If the microgrid is connected to the main power grid, the fault current flowing through CB_5 would be larger than the current flowing through CB_4. This is because the former is provided by the main grid and the latter is supported by the DGs in MG-1 [26,27]. In this case, it is vital for the protection systems of both the main power grid and MG-1 to make correct decisions and to trip the circuit breakers at the two terminals of feeder L_3. If the microgrid is operated under islanded operation, the fault current coming from the two terminals of feeder L_3, which flows through CB_4 and CB_5, is provided by the DGs of microgrids MG-1 and MG-2. Note that there are obvious differences in the fault current of the microgrid in the grid-connected mode and the islanded mode, so it is important to consider the response speed, selectivity, sensitivity and reliability of the microgrid protection configuration [28].

If the microgrid is operated in the islanded mode, the fault current flowing through CB_5 and CB_4 would also be different, because the fault current is provided by DGs from MG-2 and MG-1.

Note that significant differences in the fault current are shown when the microgrid is operated in different modes.

8.3.2 Protection at the PCC of a Microgrid

Figure 8.22 shows typical protection functions available at the point of common coupling (PCC) of a microgrid.

In order to realize the protection of MG-1, the protection and monitoring devices at the PCC, as shown in Figure 8.22, should measure the three-phase voltage at the PCC close to MG-1 (measurements from TV_M) and the three-phase current flowing through the PCC (measurements from TA_PCC). In addition, the measurements should also include

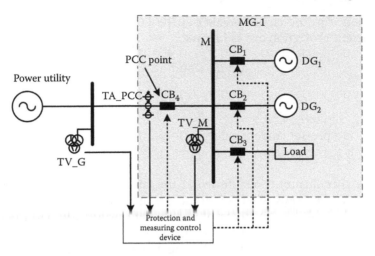

FIGURE 8.22 Diagram of protection at the PCC of MG-1.

the three-phase voltage of the main power grid (measurements from TV_G). TV_G is required by the synchronization closing control when MG-1 transforms from the islanded mode to the grid-connected mode.

8.3.2.1 Grid-Connected Mode

In the grid-connected mode, protection at the PCC (i.e. the point of connection between the microgrid and the main grid) is required to provide reverse power protection, overcurrent protection, overvoltage and undervoltage protection, over-frequency and under-frequency protection and islanding detection protection [29].

8.3.2.1.1 Reverse Power Protection

In the grid-connected mode, surplus power from the microgrid can be fed back to the main grid. This is made possible by the utility grid within a certain capacity. Implementing the reverse power protection that acts on an alarm with delay in the protecting device is convenient to notify the operating personnel about the operating state of the microgrid.

8.3.2.1.2 Overcurrent Protection

Overcurrent protection must be installed at the PCC of a microgrid. When an internal fault within the microgrid is detected, the overcurrent protection trips the circuit breaker at the PCC. The key equations of an overcurrent protection are as follows:

$$I_{max} \geq I_{set} \quad \text{and} \quad t > t_{set}, \tag{8.5}$$

where

$I_{max} = \max(I_A, I_B, I_C)$
I_{set} is a pre-set current value
t_{set} is the time delay

At some specific points of a network, the fault current might be smaller than the load current when a fault occurs in a microgrid, and this might lead to a lack of the sensitivity of overcurrent protection. On the other hand, a decrease of bus voltage occurs during the fault; so, the overcurrent protection can be activated by the detection of low voltage, which could make use of the definite-time or inverse-time characteristics [30].

8.3.2.1.3 Over/Undervoltage Protection

When the voltage at any measured point breaches the set point of the over/undervoltage protection, the protection will be enabled by tripping the circuit breakers or an alarm.

When any of the voltages, viz. V_{AB}, V_{BC} and V_{CA}, are lower than the lower set point, $V_{set.L}$, and have a certain time delay, the undervoltage protection will act. This can be shown by

$$\{[V_{AB} \leqq V_{set.L}] \text{ or } [V_{BC} \leqq V_{set.L}] \text{ or } [V_{CA} \leqq V_{set.L}]\} \text{ and } t > t_{set}, \tag{8.6}$$

When any of the voltages, i.e. V_{AB}, V_{BC} and V_{CA}, are higher than the upper set point, $V_{set.H}$, and have a certain time delay, the overvoltage protection will act. This can be shown by

$$\{[V_{AB} \geq V_{set.H}] \text{ or } [V_{BC} \geq V_{set.H}] \text{ or } [V_{CA} \geq V_{set.H}]\} \text{ and } t > t_{set}, \tag{8.7}$$

Multi-stage voltage protection can be set up according to the requirement of over/undervoltage protection.

8.3.2.1.4 Over/Under-Frequency Protection

When the frequency is lower than the lower set point, $f_{set.L}$, and has a certain time delay, the under-frequency protection will act by tripping the circuit breakers or an alarm. This can be shown by

$$f \leqq f_{set.L} \text{ and } t > t_{set}, \tag{8.8}$$

When the frequency is higher than the upper set point, $f_{\text{set.H}}$, and has a certain time delay, the over-frequency protection will act by tripping the circuit breakers or an alarm. This can be expressed as

$$f \geq f_{\text{set.H}} \text{ and } t < t_{\text{set}}, \tag{8.9}$$

Multi-stage frequency protection can be set up according to the requirement of over/under-frequency protection.

8.3.2.2 Islanded Mode

8.3.2.2.1 Voltage and Frequency at the Microgrid PCC

The electrical characteristics of a microgrid operating in an islanded mode are the basis for the detection and protection design. These characteristics include voltage magnitude-frequency, total distortion rate of voltage harmonic and voltage frequency change rate [31].

For illustration purposes, the microgrid shown in Figure 8.22 is used. Let us first assume that the microgrid is in the grid-connected mode, the sum of power output of all distributed generation is P_{DG} and Q_{DG}, the power consumed by the local load is P_{load} and Q_{load} and the power injected from the utility grid is ΔP and ΔQ. Thus, the power balance of the microgrid can be formulated by

$$\begin{cases} P_{\text{load}} = P_{\text{DG}} + \Delta P = \dfrac{V_{\text{PCC}}^2}{R}, \\[4mm] Q_{\text{load}} = Q_{\text{DG}} + \Delta Q = V_{\text{PCC}}^2 \left(\dfrac{1}{2\pi f L} - 2\pi f C \right), \end{cases} \tag{8.10}$$

where
   R, L and C are the equivalent resistance, inductance and capacitance of the power load in the microgrid
   U_{PCC} is the voltage magnitude at the PCC
   f is the system frequency

When the microgrid changes from the grid-connected to the islanded mode, its new power balance can be reformulated by

$$\begin{cases} P_{\text{load}}^* = P_{\text{DG}} = \dfrac{V_{\text{PCC}}^{*2}}{R}, \\[4mm] Q_{\text{load}}^* = Q_{\text{DG}} = V_{\text{PCC}}^{*2} \left[\dfrac{1}{2\pi(f+\Delta f)L} - 2\pi(f+\Delta f)C \right], \end{cases} \tag{8.11}$$

where

V_{PCC}^* is the voltage magnitude at the microgrid PCC

Δf is the offset between the frequency and the reference frequency of the microgrid in the islanded mode, f^* and f

According to (8.10) and (8.11), the difference in the voltage magnitude of the microgrid before and after the mode change can be expressed by

$$\Delta V_{PCC} = V_{PCC}^* - V_{PCC} = \left(\frac{V_{PCC}^*}{V_{PCC}} - 1 \right) V_{PCC} = \left(\sqrt{\frac{P_{DG}}{P_{DG} + \Delta P}} - 1 \right) V_{PCC}, \quad (8.12)$$

As can be seen from (8.12), if there is an active power exchange between the microgrid and the main grid (i.e. $\Delta P \neq 0$) in the grid-connected mode, a voltage magnitude offset will occur when the microgrid changes to the islanded mode. Similarly, if there is a reactive power exchange between the microgrid and the main grid (i.e. $\Delta Q \neq 0$) at the grid-connected mode, a voltage magnitude and frequency offset will occur when the microgrid changes to the islanded mode. Figure 8.23a shows the change in microgrid frequency when the microgrid switches from the grid-connected to the islanded mode at time $t = 0.3$ s.

8.3.2.2.2 Total Distortion Factor of Voltage Harmonic

The voltage and current output wave form of a PWM inverter can be close to the sinusoidal wave form. However, due to the use of a carrier to modulate the wave form, the output will normally carry some harmonics associated with the carrier. A Fourier decomposition of the output wave form of a PWM inverter shows that the output line-to-line voltage of a three-phase bridge PWM inverter does not contain low harmonics, but contains harmonics close to the carrier angular frequency ω_c. The harmonics with higher ω amplitude are $\omega_c \pm 2\omega_r$ and $2\omega_c \pm \omega_r$ (ω_r is the angular frequency of the modulated wave form).

In a microgrid, most DGs are inverter based, and the presence of voltage harmonics at the microgid PCC is evident. When the microgrid is in the grid-connected mode, in theory, the output of the inverter will be clamped by the utility grid and the voltage harmonic distortion factor will be relatively small. When the microgrid is in an islanded mode, because of the loss of clamping from the utility grid, the output voltage distortion of the inverter will be presented at the microgrid PCC and the distortion factor of the voltage harmonics will be relatively large.

8.3.2.2.3 Islanding Detection Methods

When a fault occurs on the utility grid side, the microgrid should be transferred to the islanded mode. Meanwhile, some DGs have to adjust their

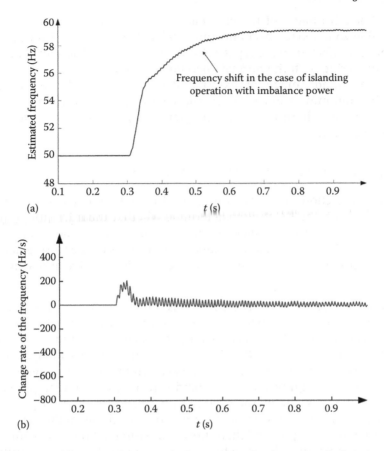

FIGURE 8.23 Change of frequency of a microgrid when switched from grid-connected to islanded mode: (a) frequency and (b) change rate of the frequency.

control strategies to ensure stable operation of the microgrid, so an action of fast and accurate islanding detection is very important. The islanding detection methods are developed to detect whether the microgrid is under the islanded condition. In general, common islanding detection methods include external detection methods and internal detection methods.

8.3.2.2.3.1 External Detection Methods

The external detection methods, also referred to as remote detection methods or grid-side detection methods, directly detect the status of the interconnection switch of the microgrid and send out signals from the grid side. The receivers installed on the DGs in the microgrid confirm whether the microgrid is in the islanded mode and further define their own control strategies [32].

The advantages of the external detection methods include a non-detection zone (NDZ), accurate and reliable detection, consistent performance not affected by the types of DG devices and grid-friendly behaviour not interfering with the normal operation of the utility grid.

The disadvantage of external detection methods is that they require communication facilities and complex configuration, which is costly, and it is hard to achieve plug-and-play or apply them to large-scale microgrids.

8.3.2.2.3.2 Internal Detection Methods

Internal detection methods, also referred to as local detection methods, mainly rely on the DG itself to determine whether the microgrid is in the islanded mode, by detecting their output voltage magnitude and network frequency [33,34] without additional transformers or measuring devices. These methods can be divided into two categories: passive and active.

1. *Passive detection methods*: These methods usually determine the microgrid operation mode by detecting irregularity occurring in the electrical characteristics of DGs, such as output voltage magnitude, frequency, phase and harmonics. Considering that the control strategies of the inverters of the DGs also need to detect terminal voltages, these methods do not require additional hardware or independent protection relays. Besides, the methods do not interfere with the utility grid and have no effect on power quality. Nevertheless, these methods have big detection blind spots and their threshold setting is difficult to configure. For example, in order to reduce the NDZ, increasing device sensitivity with lower threshold values is required, but this may cause false tripping of devices, affecting the normal operation of the system. Usually, a combination of passive and active methods is used in the case power injection or absorption is required from the main grid.

   Typical passive detection methods mainly contain over/under-voltage detection methods, over/under-frequency detection methods, voltage harmonic detection methods, sudden voltage phase methods and so on [35].

2. *Active detection methods*: These methods have been developed mainly based on the control strategy of inverter-based DGs. These methods inject small disturbances into the inverter control signal, which are consequently presented in the output voltage, frequency and power of the inverters. Since these disturbances are clamped by the microgrid power balance during the grid-connected operation, the impact of the disturbance signal on protection is not significant. When

islanding occurs, the impact of these disturbances will be presented in the form of voltage magnitude, frequency and phase deviation at the PCC. Therefore, whether the microgrid operates in the islanded mode can be determined by detecting electrical characteristics at the PCC. These methods have the advantages of small detection blind spots and high accuracy. The drawbacks are that the introduction of disturbance affects the power quality during the normal operation of a microgrid and such methods require more complicated control strategies for the inverters.

Typical active detection methods mainly contain reactance insertion methods, voltage deviation methods, frequency deviation methods, output power perturbation methods and so on [36].

8.3.3 Protection of Feeders within a Microgrid

Some protecting functions are needed when the microgrid operates in the islanded mode, which include over/undervoltage protection, over/under-frequency protection, protection based on the rate of change of frequency (ROCOF), vector shift protection, current differential protection, feeder protection, etc.

For the microgrid shown in Figure 8.22, circuit breaker CB_4 is open when MG-1 operates in the islanded mode. The over/undervoltage protection and over/under-frequency protection can be realized by measuring the bus voltage at the microgrid side of the PCC. To ensure the selectivity of protection, a current differential bus protection can be equipped for bus M. When a fault occurs at bus M, the current differential bus protection will operate to trip all the connected breakers. When a fault occurs inside the microgrid but not at bus M, over/undervoltage protection and over/under-frequency protection should operate to trip the breakers. The protection configuration for the microgrid shown in Figure 8.22 is reasonable considering that microgrid MG-1 has a simple topology; furthermore, it is acceptable that the microgrid goes out of service for the action of protection when contingencies occur in the microgrid in the islanded mode or under an abnormal condition. In the case of a large and complicated microgrid, it is necessary to configure feeder protection.

For a microgrid with a complex topology, e.g. the feeders divided into several sections, microgrid protection should have the capability to selectively clear the fault component inside a feeder section based on the microgrid's operation reliability requirement. In this case, protections should be configured for the feeder sections inside the microgrid [37].

When a microgrid operates in the islanded mode, the internal distributed generation supplies power to the load. Due to the loss of large

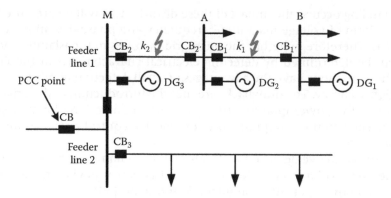

FIGURE 8.24 Schematic diagram of the protection configuration of microgrid in the islanded mode.

short-circuit capacity of the distribution power system, the fault features of internal feeders are different from those in the grid-connected mode [38–40]. Based on the microgrid structure put forward by the Consortium for Electric Reliability Technology Solutions (CERTS), this section analyses the protection configuration scheme when the microgrid operates in the islanded mode. The structure of the microgrid is shown in Figure 8.24.

In Figure 8.24, the PCC indicates the point of common connection between the microgrid and the external power grid, where the circuit breaker is open when the microgrid is in the islanded mode. Assume that all the distributed power sources are inverter-based equipment denoted by DG_i, respectively. There are two feeders in the microgrid, among which the first feeder connects the important loads which require a more reliable power supply, and the DGs are equipped in different locations on the feeder in a dispersed manner. The second feeder is not equipped with DGs. Finally, there are circuit breakers in each connection point of the distributed generation.

When the microgrid operates in the grid-connected mode, the short circuit current is large, so the circuit breakers equipped on the feeders, CB_1–CB_3, in Figure 8.24 are equipped with an instantaneous trip current protection and an overcurrent protection in order to meet the requirements for the protection scope and coordination. Definite-time and inverse-time overcurrent relays can be used. However, the short circuit current is relatively small when the microgrid is in the islanded mode, due to the nature of the inverter-based distributed generation. Therefore, voltage-controlled overcurrent protections are required in CB_1–CB_3. If the voltage-controlled overcurrent protections are still not satisfied with the requirement of sensitivity, the current-controlled undervoltage protection can be considered.

For example, when a fault occurs at point k_1 in Figure 8.24, the voltage of this point reduces significantly and there is a short circuit current flowing through CB_1. Therefore, the current-controlled undervoltage protection of CB_1 is activated to trip CB_1. However, when a fault occurs at the point k_2 in Figure 8.24, the voltage of CB_1 also reduces significantly, but the current would not be as large as that flowing through circuit breaker CB_1. At this time, the current blocking low-voltage protection would not be activated. After the occurrences of the faults, the distributed generation DG_1-DG_3 will also supply fault current. Therefore, the circuit breakers CB_1' and CB_2' should be installed at the other end of the lines, and the overcurrent protection initiated by low voltage or low-voltage protection initiated by overcurrent is implemented.

The methods discussed provide a feasible solution to the operation of the microgrid in the islanded mode. When the current protection cannot meet the requirements of the sensitivity, the current differential protection and directional pilot protection scheme can be adopted and applied directly on the feeder of the microgrid.

8.3.4 Case Studies

A simulation model of a distribution system with a microgrid is shown in Figure 8.25, where the system reference capacity is 500 MVA and the reference voltage is 10.5 kV. The maximum and minimum equivalent impedance seen from bus A to the system side are $Z_{smin} = 0.91\ \Omega$ and $Z_{smax} = 0.91\ \Omega$, respectively. Lines AB and BC are overhead lines while lines AF, DE and CD are underground cables.

8.3.4.1 Protection Setting Calculation without the Microgrids

Without considering the connection of the microgrid, the whole set of protection relays 1–5 is shown in Table 8.3.

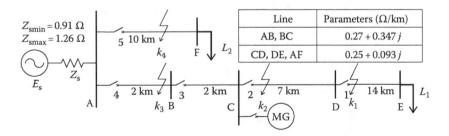

FIGURE 8.25 Simulation model of a distribution system with microgrid.

TABLE 8.3 Set of Feeder Current Protection without a Microgrid

Protection Relay	Maximum Short-Circuit Current at the Load Side of This Section (kA)	Instantaneous Current Setting Value $I_{set.I}$ (kA)	Current Setting Value $I_{set.II}$ (kA) of the Instantaneous Overcurrent Relay with Time Delay	Time Setting Value of the Delay Time Instantaneous Overcurrent Protection $t_{set.II}$ (s)
1	0.813	0.976		
2	1.667	2.0	1.037	0.5
3	3.276	3.931	2.2	0.5
4	6.295	7.554	4.324	0.5
5	2.149	2.578		

8.3.4.2 Impact of the Microgrid on Overcurrent Protections When a Fault Occurs on Its Downstream Feeders

As shown in Figure 8.25, a microgrid is connected to bus C of feeder 3.

Taking the point of connection of the microgrid as a benchmark, a downstream fault of the microgrid would be a fault occurring on the downstream feeder far away from the distribution network side, while an upstream fault of the microgrid would be a fault occurring in the upstream feeder close to the distribution network side. Here we discuss the impact of downstream faults of microgrid MG on downstream (such as relays 1 and 2) and upstream (such as relays 3 and 4) protections.

Recall that the short circuit current of inverter-based distributed generation is closely related to distributed generation types and control strategies. It is assumed that the distributed generation within the microgrid is equivalent to a constant power source, i.e. the power output of the distributed generation remains constant when a fault occurs.

8.3.4.2.1 Impact on Downstream Protection

As shown in Figure 8.25, the fault point k_1 is within the instantaneous trip current protection range, a distance of 5% of line DE. Considering the equivalent power injected by the microgrid is a given quantity denoted by S, we can obtain the impact of this quantity on the current and voltage throughout the system. The relationships between current I_{relay2} flowing through the downstream protection relay 2, current I_{MG} provided by the microgrid and power S injected by the microgrid are shown in Figure 8.26.

As we can see from Figure 8.26, after the microgrid MG is connected to bus C, the fault current through protection relay 2 increases along with the increase in the injection power S. When the microgrid injects power S that is larger than 11.17 MVA, the current flowing through protection relay 2 exceeds the fast-tripping setting value (2 kA). Thus, when a three-phase short-circuit fault occurs at point k_1, protection relays 1 and 2 will be instantly activated and will cut off the line segments of DE and CD, which means that the feeder protection loses selectivity.

8.3.4.2.2 Influence on Upstream Protection

As shown in Figure 8.26, assume that a three-phase short-circuit fault occurs at point k_2 on the downstream feeder of the microgrid, which is 25% of the total length of line CD away from bus C. Figure 8.27 shows the relationships between current I_{relay3} through the upstream protection relay 3, current I_{MG} provided by the microgrid and power S injected by the microgrid.

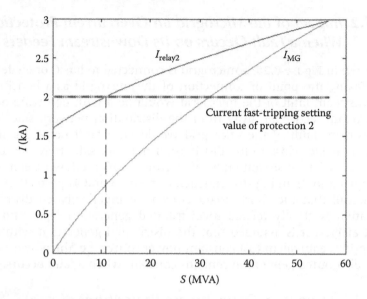

FIGURE 8.26 Relationships between I_{relay2}, I_{MG} and power S injected by the microgrid while a fault occurs at k_1.

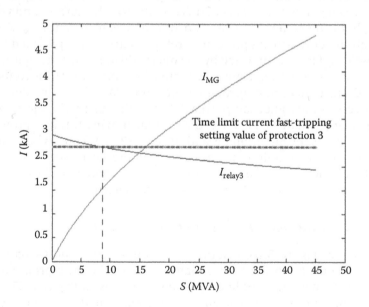

FIGURE 8.27 Relationships between I_{relay3}, I_{MG} and power S when a downstream fault occurs at k_2.

As can be seen from Figure 8.27, as the remote backup protection of protection relay 2, the time limit current fast-tripping of protection relay 3 should act when protection relay 2 cannot remove faults instantaneously. However, the fault current through protection relay 3 continues to decrease along with the increase in injection power S after the microgrid is connected to the distribution system. When the power injected by the microgrid exceeds 8.45 MVA, the fault current will be less than the time limit setting value of protection relay 3. Thus it might not be able to realize the backup protection.

In conclusion, the integration of the microgrid might increase the protection range of the downstream protections, which would cause the discrimination of the downstream protections to be lost. On the contrary, the integration of the microgrid might decrease the protection range of upstream protections and also lead to the failure of backup protection.

8.3.4.3 Impact of the Microgrid on Overcurrent Protections When a Fault Occurs on Its Upstream Feeders

When a three-phase short-circuit fault occurs at point k_3 on the upstream feeder of the microgrid MG, the current flowing through protection relay 3 is provided by the microgrid. Obviously, as shown in Figure 8.28, the

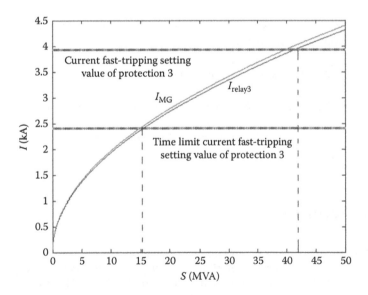

FIGURE 8.28 Relationships between I_{relay3}, I_{MG} and power S when an upstream fault occurs at k_3.

more the injection power of the microgrid, the larger the short-circuit current flowing through protection relay 3.

As can be seen from Figure 8.27, when there is a fault occurring at k_3, protection relay 3 will probably be tripped only in the case that the injection power of the microgrid is large enough. Thus protection relay 3 will not be tripped when the injection power of the microgrid is small. Furthermore, because of the short circuit current at the fault point provided by the microgrid, the electric arc at the fault point might keep burning, eventually leading to unsuccessful reclosing at the start of the feeder (protection relay 4).

8.3.4.4 Impact of the Microgrid on Overcurrent Protections When a Fault Occurs on Adjacent Feeders

As shown in Figure 8.25, when there is a fault at k_3 of feeder 2, the protection of feeder 1 will not be activated. However, after the microgrid is connected to feeder 1, the fault current provided by the microgrid will flow through protection relays 3 and 4 and may result in their misoperation.

☐ Questions

1. List the advantages of the peer-to-peer control strategy over the master-slave control strategy.

2. Explain the purposes of the outer controller and the inner controller.

3. Explain the requirements for the protection of a microgrid.

4. List the common islanding detection methods and compare their advantages and disadvantages.

☐ References

1. J. A. Pecas Lopes, C. L. Moreira, A. G. Madureira et al., Control strategies for microgrids emergency operation, *Proceedings of the International Conference on Future Power Systems*, Amsterdam, the Netherlands, 16–18 November 2005, pp. 1–6.
2. M. Milosevic and G. Andersson, Generation control in small isolated power systems, *Proceedings of the 37th Annual North American Power Symposium*, Ames, IA, 23–25 October 2005, pp. 524–529.

3. S. Cobben, Bronsbergen: The first microgrid in the Netherlands, *Proceedings of the Kythnos 2008 Symposium on Microgrids*, Kythnos Island, Greece, 2 June 2008.
4. D. Georgakis, S. Papathanassiou and N. Hatziargyriou, Operation of a prototype microgrid system based on micro-sources quipped with fast-acting power electronics interfaces, *Proceedings of the 35th Power Electronics Specialists Conference*, Aachen, Germany, 20–25 June 2004, pp. 2521–2526.
5. R. Hara, Demonstration project of 5 MW PV generator system at Wakkanai, *Proceedings of the Kythnos 2008 Symposium on Microgrids*, Kythnos Island, Greece, 2 June 2008.
6. N. Hatziargyriou, Overview of microgrid R&D in Europe, *Proceedings of the Nagoya 2007 Symposium on Microgrids*, Nogoya, Japan, 6 April 2007.
7. K. D. Brabandere, *Voltage and Frequency Droop Control in Low Voltage Grids by Distributed Generators with Inverter Front-End*. Leuven, Belgium: Katholieke University, 2006.
8. A. Timbus, R. Teodorescu, F. Blaabjerg et al., Linear and nonlinear control of distributed power generation systems, *Proceedings of 2006 IEEE Industry Applications Conference*, Tampa, FL, 8–12 October 2006, pp. 1015–1023.
9. Z. Huibin, B. Arnet, L. Haines et al., Grid synchronization control without AC voltage sensors, *Proceedings of the 18th IEEE Applied Power Electronics Conference and Exposition*, Miami Beach, FL, 9–13 February 2003, pp. 172–178.
10. B. Frede, T. Remus, L. Marco et al., Overview of control and grid synchronization for distributed power generation systems, *IEEE Transactions on Industrial Electronics*, 53(5), 1398–1409, 2006.
11. S. K. Kim, J. H. Jeon, C. H. Cho et al., Modeling and simulation of a grid-connected PV generation system for electromagnetic transient analysis, *Solar Energy*, 83(5), 664–678, 2009.
12. C. Wang, M. H. Nehrir and H. Gao, Control of PEM fuel cell distributed generation systems, *IEEE Transactions on Energy Conversion*, 21(2), 586–595, 2006.
13. C. Wang, *Modeling and Control of Hybrid Wind Photovoltaic Fuel Cell Distributed Generation Systems*. Bozeman, MT: Montana State University, 2006.
14. A. Bertani, C. Bossi, F. Fornari et al., A microturbine generation system for grid connected and islanding operation, *Proceedings of 2004 IEEE PES Power Systems Conference and Exposition*, New York, 10–13 October 2004, pp. 360–365.
15. K. Brabandere, K. Vanthournout, J. Driesen et al., Control of microgrids, *Proceedings of 2007 IEEE Power Engineering Society General Meeting*, Tampa, FL, 24–28 June 2007, pp. 1–7.
16. P. Karlsson, J. Björnstedt and M. Ström, Stability of voltage and frequency control in distributed generation based on parallel-connected converters feeding constant power loads, *IEEE European Conference on Power Electronic and Applications*, September 2005, Dresden, Germany, pp. 1–10.
17. U. Borup, F. Blaabjerg and P. Enjeti, Sharing of nonlinear load in parallel-connected three-phase converters, *IEEE Transactions on Industry Applications*, 37(6), 1817–1823, 2001.
18. P. Piagi and R. Lasseter, Autonomous control of microgrids, *Proceedings of 2006 IEEE Power Engineering Society General Meeting*, Piscataway, NJ, 18–22 June 2006, pp. 1–8.

19. J. Guerrero, L. Vicuna, J. Matas et al., A wireless controller to enhance dynamic performance of parallel inverters in distributed generation system, *IEEE Transactions on Industrial Electronics*, 19(5), 1205–1213, 2004.

20. H. Laaksonen, P. Saari and R. Komulainen, Voltage and frequency control of inverter based weak LV network microgrid, *Proceedings of 2005 International Conference on Future Power Systems*, Amsterdam, the Netherlands, 16–18 November 2005, pp. 1263–1270.

21. T. Erika and D. Holmes, Grid current regulation of a three-phase voltage source inverter with an LCL input filter, *IEEE Transactions on Power Electronics*, 18(3), 888–895, 2003.

22. T. Hornik and Q. Zhong, Control of grid-connected DC–AC converters in distributed generation experimental comparison of different schemes, *Compatibility and Power Electronics*, 20(22), 271–278, 2009.

23. A. Timbus, M. Liserre, R. Teodorescu et al., Evaluation of current controllers for distributed power generation systems, *IEEE Transactions on Power Electronics*, 24(3), 654–664, 2009.

24. J. Dannehl, C. Wessels and F. W. Fuchs, Limitations of voltage-oriented PI current control of grid-connected PWM rectifiers with filters, *IEEE Transactions on Industrial Electronics*, 56(2), 380–388, 2009.

25. IEEE Std 1547.4-2011, Guide for design, operation, and integration of distributed resource island systems with electric power systems, IEEE Standards Coordinating Committee 21 on Fuel Cells, Photovoltaics, Dispersed Generation and Energy Storage, New York, 20 July 2011.

26. A. Abdel-Khalik, A. Elserougi, A. Massoud et al., Fault current contribution of medium voltage inverter and doubly-fed induction machine-based flywheel energy storage system, *IEEE Transactions on Sustainable Energy*, 4(1), 58–67, 2013.

27. N. Nimpitiwan, G. Heydt, R. Ayyanar and S. Suryanarayana, Fault current contribution from synchronous machine and inverter based distributed generators, *IEEE Transactions on Power Delivery*, 22(1), 634–641, 2007.

28. M.A. Haj-ahmed and M. S. Illindala, The influence of inverter-based DGs and their controllers on distribution network protection, *IEEE Transactions on Industry Applications*, 50(4), 2928–2937, 2013.

29. N. Jenkins, R. Allan, P. Crossley, D. Kirschen and G. Strbac, *Embedded Generation*. London, UK: The Institution of Engineering and Technology, 2008.

30. S. H. Horowitz and A. G. Phadke, *Power System Relaying*. Chichester, UK: John Wiley & Sons Ltd., 2008.

31. B. Li, Z. Bo, X. Yu et al., Protection schemes for closed loop distribution network with distributed generator, *The First International Conference on Sustainable Power Generation and Supply*, Nanjing, China, 6–7 April 2009.

32. M. A. Zamani, T. S. Sidhu and A. Yazdani, A protection strategy and microprocessor-based relay for low-voltage microgrids, *IEEE Transactions on Power Delivery*, 26(3), 1873–1883, 2011.

33. H. Wan, K. K. Li and K. P. Wong, An adaptive multi-agent approach to protection relay coordination with distributed generators in industrial power distribution system, *IEEE Transactions on Industry Applications*, 46(5), 2118–2124, 2010.

34. S. M. Brahma and A. A. Girgis, Development of adaptive protection scheme for distribution systems with high penetration of distributed generation, *IEEE Transactions on Power Delivery*, 19(1), 56–63, 2004.
35. S. I. Jang and K. H. Kim, An islanding detection method for distributed generations using voltage unbalance and total harmonic distortion of current, *IEEE Transactions on Power Delivery*, 19(2), 745–752, 2004.
36. M. A. Referrn, O. Usta and G. Fielding, Protection against loss of utility grid supply for a dispersed storage and generation unit, *IEEE Transactions on Power Delivery*, 8(3), 948–954, 1993.
37. Z. Ye, A. Kolwalkar, Y. Zhang et al., Evaluation of anti-islanding schemes based on nondetection zone concept, *IEEE Transactions on Power Electronics*, 19(5), 1171–1176, 2004.
38. D. M. Francesco, L. Marco and P. Alberto, Overview of anti-islanding algorithms for PV systems. Part 1: Passive methods, *Proceedings of the 12th International Conference on Power Electronics and Motion Control*, Portoroz, Slovenia, 2007, pp. 1878–1883.
39. B. Li, J. Wang, H. Bao and H. Zhang, Islanding detection for microgrid based on frequency tracking using extended Kalman filter algorithm, *Journal of Applied Mathematics*, 2014(10), 1–11 2014.
40. X. Chen and Y. Li, An islanding detection algorithm for inver-based distributed generation based on reactive power control, *IEEE Transactions on Power Electronics*, 3(73), 1–12, 2013.

CHAPTER

9

Energy Management and Optimal Planning of Microgrids

9.1 Introduction

Energy management plays an important role in the safe, reliable and economical operation of microgrids. Energy management is responsible for scheduling and controlling the distributed generators, energy storage devices and controllable loads based on the operating conditions of the system. The optimal planning of microgrids is also crucial, which aims to obtain the best solution to system configuration (e.g. the types and capacities of the devices) according to energy demand and available distributed energy resources during the planning period, together with the specified planning objectives and system constraints. Due to the need for an economic analysis of the devices during their life cycle, the selection of proper operational strategies for microgrids becomes an important component of optimal planning, which means that the operational strategies will greatly affect the final planning solution. Energy management and optimal planning of microgrids are the focus of this chapter.

9.2 Energy Management of Microgrids

9.2.1 Microgrid Energy Management System

A microgrid energy management system (MEMS) contains two main modules, which are supervisory control and data acquisition (SCADA) and energy management, as well as the associated hardware and software systems to support the functions of these two modules.

The main aim of MEMS is to achieve coordinated control and optimal operation of microgrids, through system-level management and control.

9.2.1.1 SCADA

In microgrids, the main function of SCADA is to transmit real-time monitoring data to the database and to the energy management module, as well as to convey the generation scheduling and control instructions from the energy management module to the local control systems. In microgrids, distributed generators are equipped with local control systems, and the SCADA system bridges the energy management module and these local control systems.

9.2.1.2 Energy Management Module

The energy management module is the core of MEMS. Based on real-time and historical data, as well as the policy and market information, etc., the energy management module is able to produce operational scheduling for various distributed generators, energy storage devices and the controllable loads in a microgrid. These operational schedules are sent to these local devices via the SCADA system. From the functionality point of view, the energy management module can be divided into four parts:

1. *Data forecasting*: Forecasting demand and output of distributed generators [1]

2. *Optimal scheduling*: Formulating scheduling plans based on the economic, environmental and technical requirements and the microgrid operational constraints [2]

3. *Operation control*: Real-time or near-real-time monitoring and adjusting the operation of the devices to maintain system voltage and frequency stability [3]

4. *Data analysis*: Analysis of all the real-time and historical data in order to obtain information related to system economics, stability and security

9.2.2 Microgrid Scheduling Strategy

The microgrid scheduling strategy is mainly used to schedule the power output of distributed generators and energy storage systems. A reasonable scheduling strategy can enable the microgrid to achieve the predetermined operation goals, such as minimum operation cost, maximum efficiency of system energy utilization and lowest carbon emissions. Microgrid scheduling strategies can be divided into two categories: heuristic scheduling strategies [4–6] and optimization strategies [7]. Optimization strategies can be further divided into static optimization and dynamic optimization.

9.2.2.1 Heuristic Scheduling Strategy

In theory, wind, photovoltaics (PV) and other renewable energy sources should prioritize meeting the demand. Based on this, the heuristic scheduling strategy formulates the generation priorities of different distributed generators associated with the start/stop conditions of conventional distributed generators (e.g. diesel generation units) and the charging/discharging conditions of energy storage systems. Such scheduling ensures effective operation of these devices.

9.2.2.2 Static Optimization Strategy

Based on the availability of distributed generation and their operational cost per unit (e.g. per kWh) during a period, as well as the requirements of demand at the same period, the static optimization strategy produces the optimal dispatch values of all devices (including the distributed generators and energy storage devices) [8]. This strategy usually does not consider the correlation of the available resources among different time periods, and thus the optimization is often formulated for each time period independently. In order to achieve the most cost-effective operation, the availability and costs of all schedulable power sources should be considered. For a microgrid operating in grid-connected mode, the power from the main grid should also be considered as a schedulable source, participating in the cost comparison with other distributed energy resources.

9.2.2.3 *Dynamic Optimization Strategy*

Due to the relatively low ramp rates of some distributed generators (e.g. diesel generators, micro-turbines) and the constraint of the residual capacity of energy storage systems from a previous time period, only considering the available resources at the current time period sometimes will not result in the optimal operation scheduling. Therefore, dynamic optimization strategies have been proposed to optimize system operation with the goal of maximizing the overall benefits for the scheduling period [9], taking into account the predictive data of multiple time periods. Due to the daily cycle characteristics of demand and output of distributed generators (e.g. PV), dynamic optimization strategies are usually implemented daily to develop the solution of the so-called day-ahead dynamic economic scheduling. In order to reduce the coupling of different cycles, the residual capacities of energy storage systems are commonly required to be the same at the start and end of the scheduling cycles. In addition, the prediction is usually updated by using the recently monitored data to increase the accuracy, based on which the optimal schedules can be improved in a rolling manner.

Heuristic scheduling strategies usually need to be formulated considering the specific network configurations, and thus require to be tuned for each network, making it less practical to be implemented. This chapter will focus on the optimization scheduling strategies.

9.2.3 Microgrid Device Models

Depending on the focuses of various applications, models of the devices that are used in a microgrid can be different. For the optimal scheduling of a microgrid, steady-state models reflecting the relations between inputs and outputs of the devices are usually used. Since there is no standard configuration for a microgrid, and microgrids can contain various types of devices, a network model of a microgrids is normally tuned according to user requirements. In order to facilitate the elaboration, the microgrid shown in Figure 9.1 is taken as an example to introduce the models of the devices used.

The microgrid shown in Figure 9.1 consists of a PV system, a micro-turbine, a battery energy storage system, a gas boiler, a waste heat recovery boiler, an electric chiller and a steam/water heat exchanger. This system is capable of meeting the user demand for electricity, cooling and hot water.

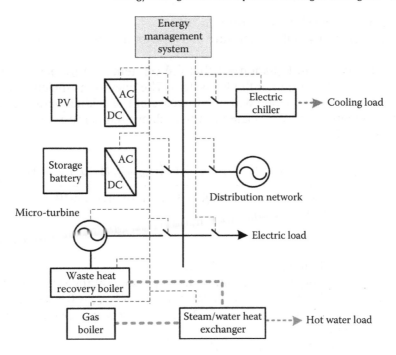

FIGURE 9.1 Structure of a microgrid.

9.2.3.1 PV Generation System

The power output of a PV generation system is closely related to the solar radiation irradiated on the surface of PV arrays, the operation conditions and physical parameters of the system, etc. Normally, the output characteristics of the PV system can be described as [8]

$$P_{PV} = f_{PV}P_{PV,R}\left(\frac{G_T}{G_{T,STC}}\right)\left[1 + \alpha_p\left(T_{cell} - T_{cell,STC}\right)\right],\qquad(9.1)$$

where

P_{PV} represents the power output of the PV system (in kW)

f_{PV} is the derating factor, reflecting the influences of surface dirt, ageing and other factors on the power output

$P_{PV,R}$ is the rated capacity of the PV system (in kW)

G_T is the actual solar radiation (in kW/m²)

$G_{T,STC}$ is the solar radiation at standard test condition (STC) (in kW/m²), which generally takes the value of 1 kW/m²

α_p is the power temperature coefficient

T_{cell} is the PV cell temperature (in °C)

$T_{cell,STC}$ is the PV cell temperature at STC with a value of 25 (in °C)

It can be seen from Equation 9.1 that the cell temperature has a certain influence on the output of the PV system. The operating efficiency of the PV system decreases along with the increase of the cell temperature. The cell temperature of PV is related to the ambient temperature, which can be calculated by

$$T_{cell} = \frac{T_a + \left(T_{cell,NOCT} - T_{a,NOCT}\right)\left(G_T/G_{T,NOCT}\right)\left[1 - \left(\left(\eta_{MPP,STC}\left(1 - \alpha_p T_{cell,STC}\right)\right)/\tau\alpha_s\right)\right]}{1 + \left(T_{cell,NOCT} - T_{a,NOCT}\right)\left(G_T/G_{T,NOCT}\right)\left(\alpha_p \eta_{MPP,STC}/\tau\alpha_s\right)},$$

$$(9.2)$$

where

T_a is the ambient temperature (in °C)

$T_{cell,NOCT}$ is the nominal operating cell temperature at the standard operating condition (where the solar radiation $G_{T,NOCT}$ is 0.8 kW/m², the ambient temperature $T_{a,NOCT}$ is 20°C, and the wind speed is 1 m/s), with a range from 45°C to 48°C

$\eta_{MPP,STC}$ is the PV efficiency at STC

τ is the solar transmittance of any cover over the PV arrays

α_s is the solar absorptance of the PV system

The default value of $\tau\alpha_s$ can be set as 0.9.

9.2.3.2 Micro-Turbine Power Generation System

The fuel consumption of a micro-turbine power generation system is usually represented by a quadratic function of its power output. For simplicity, we can also adopt a linear function to describe the relationship between the fuel consumption and output power of the micro-turbine. For example, the natural gas consumption F_{gen} (in m³) during Δt (h) can be expressed as

$$F_{gen} = u_{gen}\left(F_0 P_{gen,R} + F_1 P_{gen}\right)\Delta t, \qquad (9.3)$$

where

F_0 and F_1 are the intercept coefficient and slope of the fuel curve (both in m³/kWh)

$P_{gen,R}$ and P_{gen} are the rated power and actual power output of the turbine (both in kW)

u_{gen} is the operating state of the power generation system

If the micro-turbine is shut down, u_{gen} takes the value of 0, otherwise the value is 1.

Equation 9.3 can be further simplified to the following expression by neglecting the fuel consumption of the micro-turbine under no-load condition:

$$F_{gen} = u_{gen} \frac{P_{gen} \Delta t}{\gamma \eta_{gen}},$$ (9.4)

where
 γ refers to a parameter of unit conversion of the natural gas consumption
 η_{gen} represents the thermal efficiency of the micro turbine

The micro-turbine power generation system should meet certain operation constraints, such as the maximum and the minimum power outputs, the maximum ramp rate and the minimum run time [10], among which the power output constraint can be expressed as

$$P_{gen}^{min} \leq P_{gen} \leq P_{gen}^{max},$$ (9.5)

where P_{gen}^{min} and P_{gen}^{max} are the minimum and the maximum power output of the generator (both in kW).

The waste heat generated by the power generation system can be recycled when the system operates as a cogeneration unit. Assume that the system has a fixed heat to power ratio, which means that a fixed amount of thermal energy is produced along with one unit electric power generated by the cogeneration unit. The thermal energy recycled by the cogeneration unit can be expressed as follows:

$$Q_{gen} = \alpha_{gen} P_{gen},$$ (9.6)

where
 Q_{gen} is the heat power output of the cogeneration unit (in kW)
 α_{gen} is the heat to power ratio of the system

When considering operation optimization of the micro-turbine cogeneration system, an inaccurate equipment model sometimes cannot meet the requirements of the optimization of multiple operating conditions, and hence may reduce the accuracy and practical value of the optimization results.

9.2.3.3 Battery Energy Storage

The batteries are one type of electric energy storage system (ESS). The stored energy of an ESS, usually defined as state of charge (SOC), will change as the battery is charged or discharged. There are different types of ESSs, all of which can be modelled using simplified linear expressions. Assuming that the charge and discharge power remain constant within the time step Δt, the relation of stored energy of an ESS before and after the charge/discharge process can be described as

$$W_{\text{ESS}} = W'_{\text{ESS}}\left(1 - \sigma_{\text{ESS}}\Delta t\right) + \left(P_{\text{ESS,C}}\eta_{\text{ESS,C}} - \frac{P_{\text{ESS,D}}}{\eta_{\text{ESS,D}}}\right)\Delta t, \qquad (9.7)$$

where

$\quad W'_{\text{ESS}}$ and W_{ESS} are the stored energy of the ESS at the start and end of time step Δt, respectively (both in kWh)

$\quad \sigma_{\text{ESS}}$ is the hourly self-discharge rate of the ESS (h^{-1})

$\quad P_{\text{ESS,C}}$ and $P_{\text{ESS,D}}$ are the charge and discharge power of the ESS (both in kW)

$\quad \eta_{\text{ESS,C}}$ and $\eta_{\text{ESS,D}}$ are the charging and discharging efficiency of the ESS

The SOC of the ESS can be estimated as

$$S_{\text{ESS}} = \frac{W_{\text{ESS}}}{W_{\text{ESS,R}}}, \qquad (9.8)$$

where

$\quad S_{\text{ESS}}$ is the SOC of the ESS

$\quad W_{\text{ESS,R}}$ is the rated capacity of the ESS (in kWh)

In practice, the SOC of the ESS is limited within a range, and the charge and discharge power constraints of the ESS should be taken into consideration as well. These constraints can be expressed as

$$W_{\text{ESS}}^{\min} \leq W_{\text{ESS}} \leq W_{\text{ESS}}^{\max}, \qquad (9.9)$$

$$0 \leq P_{\text{ESS,C}} \leq P_{\text{ESS,C}}^{\max}, \qquad (9.10)$$

$$0 \leq P_{\text{ESS,D}} \leq P_{\text{ESS,D}}^{\max}, \qquad (9.11)$$

where

$\quad W_{\text{ESS}}^{\min}$ and $W_{\text{ESS}}^{\max}$ are the minimum and the maximum stored energy of the ESS (both in kWh)

$\quad P_{\text{ESS,C}}^{\max}$ and $P_{\text{ESS,D}}^{\max}$ are the maximum charge and discharge power of the ESS (both in kW)

Considering that the energy storage system cannot be in the charge and discharge state at the same time, a variable u_{ESS} is introduced. This variable will take the value of 0 if the ESS is in the charge state, otherwise it will be set as 1, indicating that the ESS is in the discharge state. Furthermore, Equations 9.10 and 9.11 can be modified as

$$0 \le P_{\text{ESS,C}} \le \left(1 - u_{\text{ESS}}\right) P_{\text{ESS,C}}^{\max}, \tag{9.12}$$

$$0 \le P_{\text{ESS,D}} \le u_{\text{ESS}} P_{\text{ESS,D}}^{\max}. \tag{9.13}$$

Additionally, for simplicity, the lifetime of the ESS is assumed to be independent of the discharge depths. Therefore, only the total through-put needs to be considered when calculating the lifetime of the ESS. Consequently, the lifetime of the ESS can be estimated by the following expression:

$$l_{\text{ESS}} = \min\left(\frac{E_{\text{ESS}}^{\text{lifetime}}}{E_{\text{ESS}}^{\text{ann}}}, l_{\text{ESS,f}} \right), \tag{9.14}$$

where

l_{ESS} is the lifetime of the ESS (years)

$E_{\text{ESS}}^{\text{lifetime}}$ is the lifetime throughput of the ESS (kWh), corresponding to the maximum amount of energy that can be cycled through the ESS before it needs to be replaced

$E_{\text{ESS}}^{\text{ann}}$ is the annual throughput (kWh)

$l_{\text{ESS,f}}$ is its float life (years), referring to the maximum service life of the ESS before it requires replacement due to its natural corrosion processes

9.2.3.4 Power Electronic Converter

Power electronic converters are required to be the interface of convert-ing DC to AC power. The main parameters of converters are capacity and energy conversion efficiency. For a bidirectional converter that acts as a rectifier or inverter, the capacity and efficiency need to be considered independently, which can be expressed by

$$P_{\text{con,AC}}^{\text{out}} = \eta_{\text{inv}} P_{\text{con,DC}}^{\text{in}},$$
$$P_{\text{con,DC}}^{\text{out}} = \eta_{\text{rec}} P_{\text{con,AC}}^{\text{in}}, \tag{9.15}$$

where

$P_{\text{con,DC}}^{\text{in}}$ and $P_{\text{con,DC}}^{\text{out}}$ are the power input and output of the bidirectional converter on the rectifier side (DC side) (both in kW)

$P_{\text{con,AC}}^{\text{in}}$ and $P_{\text{con,AC}}^{\text{out}}$ are the power input and output of the bidirectional converter on the inverter side (AC side) (both in kW)

η_{rec} and η_{inv} are the rectifier and inverter efficiency of the bidirectional converter

In addition, the power inputs of the rectifier side and the inverter side should remain within certain ranges, as follows:

$$0 \leq P_{con,DC}^{in} \leq P_{inv,R},$$
$$0 \leq P_{con,AC}^{in} \leq P_{rec,R},$$
(9.16)

where $P_{rec,R}$ and $P_{inv,R}$ are the rated active power inputs of the bidirectional converter under the rectifier and inverter states (both in kW).

The model of unidirectional rectifier or inverter can be deduced according to the equality constraints (9.15) and inequality constraints (9.16). Taking the unidirectional inverter as an example, the model can be expressed by

$$P_{inv,AC}^{out} = \eta_{inv} P_{inv,DC}^{in},$$
$$0 \leq P_{inv,DC}^{in} \leq P_{inv,R}.$$
(9.17)

9.2.3.5 Gas Boiler

When there is not sufficient thermal energy supplied by the microgrid, gas boilers can be used. Assuming that the gas boilers convert the chemical energy stored in natural gas into thermal energy with a constant efficiency, the fuel consumption of the gas boiler during Δt can be calculated by

$$F_{GB} = \gamma \frac{Q_{GB,steam}\Delta t}{\eta_{GB}},$$
(9.18)

where

F_{GB} is the fuel consumption of the gas boiler (m³)
$Q_{GB,steam}$ is the thermal power output of the gas boiler (kW)
η_{GB} is the efficiency of the gas boiler
γ is the conversion coefficient from gas to thermal energy (m³/kWh)

Meanwhile, the thermal power output of the gas boiler is limited by its rated power, which is expressed by

$$0 \leq Q_{GB,steam} \leq Q_{GB,R},$$
(9.19)

where $Q_{GB,R}$ is the rated thermal power output of the gas boiler (kW).

9.2.3.6 Waste Heat Recovery Boiler

A waste heat recovery boiler can produce hot water or steam by recovering heat in the waste gas of the generators. The energy conversion relationship can be described as

$$Q_{HR,steam}^{out} = \eta_{HR} Q_{HR,smoke}^{in}, \tag{9.20}$$

where
$Q_{HR,steam}^{out}$ and $Q_{HR,smoke}^{in}$ are the heat power output and thermal power of smoke input of the waste heat recovery boiler (both in kW)

η_{HR} is the efficiency of the waste heat recovery boiler

Meanwhile, the thermal power input of the waste heat recovery boiler is limited by the rated power, which can be formulated as

$$0 \le Q_{HR,smoke}^{in} \le Q_{HR,R}, \tag{9.21}$$

where $Q_{HR,R}$ is the rated thermal power input of the waste heat recovery boiler (in kW).

9.2.3.7 Heat Exchanger

A heat exchanger is a device that transfers heat from one heat-carrying medium to another. Heat exchangers can be exhaust gas/water heat exchangers, steam/water heat exchangers, etc. based on the need. There are some losses in the heat transfer process, and the general form of energy conversion of the heat exchanger can be expressed by

$$Q_{HX,water}^{out} = \eta_{HX} Q_{HX,steam}^{in}, \tag{9.22}$$

where
$Q_{HX,water}^{out}$ is the thermal power output of the heat exchanger (in kW)
$Q_{HX,steam}^{in}$ is the thermal power input of the heat exchanger (in kW)
η_{HX} is the conversion efficiency of the heat exchanger (the conversion efficiency is different for diverse heat exchangers)

Meanwhile, the thermal power input of a heat exchanger is limited by its rated power, and this can expressed by

$$0 \le Q_{HX,steam}^{in} \le Q_{HX,R}, \tag{9.23}$$

where $Q_{HX,R}$ is the rated thermal power input of the heat exchanger (in kW).

9.2.3.8 Electric Chiller

An electric chiller produces cooling energy by utilizing electrical energy and the relationship between them can be expressed by

$$Q_{EC,cooling}^{out} = COP_{EC}P_{EC}^{in}, \tag{9.24}$$

where
$Q_{EC,cooling}^{out}$ and P_{EC}^{in} are the cooling power output and electric power input of the electric chiller (both in kW)
COP_{EC} is the coefficient of performance of the electric chiller

Additionally, the electric power input of the electric chiller is limited by the rated power, which is expressed by

$$0 \leq P_{EC}^{in} \leq P_{EC,R}, \tag{9.25}$$

where $P_{EC,R}$ is the rated electric power input of the electric chiller (in kW).

9.2.4 Energy Balance

The energy flow of the electricity/fuel/cooling/heat supply systems of the microgrid shown in Figure 9.1 is presented in Figure 9.2.

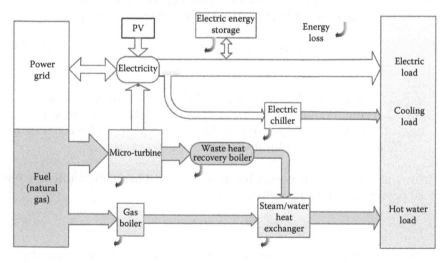

FIGURE 9.2 Energy flow in a microgrid.

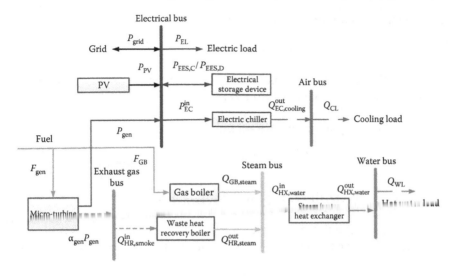

FIGURE 9.3 Bus-based energy balance structure of a microgrid.

The components of the system shown in Figure 9.2 can be divided into four categories: sources, conversion devices, energy storage and loads. From the energy delivery medium point of view, the components can also be divided into electricity, exhaust gas (waste heat of the micro-turbine), steam (output of waste heat recovery boiler and gas boiler), air (output of electric chiller) and hot water. Based on the balance relationships of these mediums, the energy flow can be described by a bus-based energy balance structure, which is presented in Figure 9.3. In this figure, the buses represent the points at which the same energy delivery mediums converge.

In accordance with the system components and structure shown in Figure 9.3, the balance expression of energy flow for each device in the system can be presented.

1. The balance equation of electric bus

$$P_{grid} + P_{gen} + P_{PV} + P_{EES,D} = P_{EL} + P_{EC}^{in} + P_{EES,C}, \tag{9.26}$$

where

P_{grid} is the electric power exchange from the grid (in kW), which is positive when the microgrid consumes electricity from the grid and negative when the microgrid supplies electricity to the grid

P_{EL} is the electric load of the system (in kW)

P_{EC}^{in} is the electric power input of the electric chiller (in kW)

2. The balance equation of an exhaust gas bus

$$\alpha_{gen} P_{gen} = Q_{HR,smoke}^{in}, \tag{9.27}$$

where $Q_{HR,smoke}^{in}$ is the thermal power of the exhaust gas absorbed by the waste heat recovery boiler (in kW).

3. The balance equation of a steam bus

$$Q_{HR,steam}^{out} + Q_{GB,steam} = Q_{HX,steam}^{in}, \tag{9.28}$$

where
$Q_{HR,steam}^{out}$ is the thermal power output of the waste heat recovery boiler (in kW)
$Q_{GB,steam}$ is the thermal power output of the gas boiler (in kW)
$Q_{HX,steam}^{in}$ is the power input of the steam/water heat exchanger (in kW)

4. The balance equation of an air bus

$$Q_{EC,cooling}^{out} = Q_{CL}, \tag{9.29}$$

where
$Q_{EC,cooling}^{out}$ is the cooling power output of the electric chiller (in kW)
Q_{CL} is the system cooling load (in kW)

5. The balance equation of the water bus

$$Q_{HX,water}^{out} = Q_{WL}, \tag{9.30}$$

where
$Q_{HX,water}^{out}$ is the power output of the steam/water heat exchanger (in kW)
Q_{WL} is the hot water load (in kW)

It is important to highlight that the bus-based energy balance structure shown in Figure 9.3 can be of great help in describing the balancing of various energy mediums in a large microgrid.

9.2.5 Optimal Scheduling Model of a Microgrid

The optimal scheduling of a microgrid is carried out in this section, taking day-ahead dynamic economic scheduling with a time step of 1 h as an example. The microgrid shown in Figure 9.1 and the corresponding steady-state models of the devices introduced previously (including PV, micro-turbine generation system, battery energy storage, power electronic converter, gas boiler, waste heat recovery boiler, heat exchanger and electric chiller) are employed here.

9.2.5.1 Objective Function

The objective function is to minimize the cost of system operation, which is expressed by

$$\min f = C_{\text{grid}} + C_{\text{fuel}}, \qquad (9.31)$$

where
C_{fuel} is the cost of the natural gas (in $)
C_{grid} is the cost of the daily total electricity purchase from the main grid (in $)
C_{grid} can be calculated by

$$C_{\text{grid}} = \sum_{n=1}^{24} (c_{\text{grid}}^n P_{\text{grid}}^n \Delta t), \qquad (9.32)$$

where
P_{grid}^n represents the purchased/sold power from/to the grid at time step n (in kW)
c_{grid}^n is the purchasing/selling price at time step n (in $/kWh)
Δt is the time step, equal to 1 h in this case

The purchasing cost of natural gas can be obtained by

$$C_{\text{fuel}} = c_{\text{fuel}} \sum_{n=1}^{24} \left(F_{\text{gen}}^n + F_{\text{GB}}^n \right)$$

$$= c_{\text{fuel}} \sum_{n=1}^{24} \left(\gamma \frac{P_{\text{gen}}^n \Delta t}{\eta_{\text{gen}}} + \gamma \frac{Q_{\text{GB}}^n \Delta t}{\eta_{\text{GB}}} \right), \qquad (9.33)$$

where
F_{gen}^n and F_{GB}^n represent the amounts of natural gas consumed by the gas turbine and gas boiler at time step n (in m³)
P_{gen}^n and Q_{GB}^n refer to the electrical power of gas turbine and the thermal power of gas boiler at time step n (in kW)
c_{fuel} is the purchasing price of natural gas per unit volume ($/m³)

9.2.5.2 Constraints

Based on the bus energy balance structure shown in Figure 9.3, the constraints include the bus power balance equations, the operating limits of devices in the microgrid and restrictions related to tie-lines between the microgrid and the main grid. Specifically, for a specific time step n, the optimal dispatching model should meet the constraints of Equations 9.26 through 9.30. Meanwhile, it should also be limited by the equality and inequality constraints related to the devices given by Equations 9.1 through 9.25.

For the battery energy storage, the residual capacity is usually assumed to be equal at the start and end of the scheduling cycles, which is expressed by

$$W_{ESS}^0 = W_{ESS}^{24},\tag{9.34}$$

where W_{ESS}^0 and W_{ESS}^{24} represent the residual capacity of ESS at the start and end of the scheduling cycles (both in kWh).

The power constraint of the tie-line between the microgrid and the main grid is expressed by

$$P_{line}^{min} \le P_{grid}^n \le P_{line}^{max},\tag{9.35}$$

where P_{line}^{min} and P_{line}^{max} are the lower and the upper power limits of the tie-line (both in kW), and P_{line}^{min} can be a negative value representing the maximum reverse power flow from the microgrid to the main grid (in kW)

9.2.5.3 Optimization Variables

The variables of the microgrid optimal scheduling model are primarily power outputs of different sources (including distributed generators and energy storage devices) and the power exchanges between the microgrid and the grid. For the microgrid shown in Figure 9.1, the variables can be expressed by

$$X^n = \left[P_{gen}^n, Q_{GB,steam}^n, P_{ESS,C}^n, P_{ESS,D}^n, P_{grid}^n, P_{EC}^{in,n}, Q_{HX,steam}^{in,n}, Q_{HR,smoke}^{in,n}, W_{ESS}^n, u_{gen}^n, u_{ESS}^n \right]^T.\tag{9.36}$$

Assume that the minimum power output of the micro-turbine is not less than 30% of its rated power, and the SOC of the battery storage system is limited within the range of 20%–90% of its rated capacity. Consequently, according to this analysis, the day-ahead dynamic optimal scheduling model of the microgrid shown in Figure 9.1 can be expressed as

$$\min\left[\sum_{n=1}^{24} c_{\text{grid}}^n P_{\text{grid}}^n + c_{\text{fuel}}\sum_{n=1}^{24}\left(\gamma\frac{P_{\text{gen}}^n}{\eta_{\text{gen}}} + \gamma\frac{Q_{\text{GB,steam}}^n}{\eta_{\text{GB}}}\right)\right],$$

s.t.

$$\begin{cases}
P_{\text{grid}}^n + P_{\text{gen}}^n - P_{\text{ESS,C}}^n + P_{\text{ESS,D}}^n - P_{\text{EC}}^{\text{in},n} = P_{\text{EL}}^n - P_{\text{PV}}^n, \\[4pt]
Q_{\text{HR,smoke}}^{\text{in},n} = \alpha_{\text{gen}} P_{\text{gen}}^n, \\[4pt]
\eta_{\text{HR}} Q_{\text{HR,smoke}}^{\text{in},n} + Q_{\text{GB,steam}}^n - Q_{\text{HX,steam}}^{\text{in},n} = 0, \\[4pt]
\eta_{\text{HX}} Q_{\text{HX,steam}}^{\text{in},n} = Q_{\text{WI}}^n, \\[4pt]
\text{COP}_{\text{EC}} P_{\text{EC}}^{\text{in},n} = Q_{\text{CL}}^n, \\[4pt]
0.3 u_{\text{gen}}^n P_{\text{gen,R}} \leq P_{\text{gen}}^n \leq u_{\text{gen}}^n P_{\text{gen,R}}, \\[4pt]
0 \leq Q_{\text{GB,steam}}^n \leq Q_{\text{GB,R}}, \\[4pt]
W_{\text{ESS}}^n = W_{\text{ESS}}^{n-1}\left(1-\sigma_{\text{ESS}}\right) + \left(P_{\text{ESS,C}}^n \eta_{\text{ESS,C}} - \dfrac{P_{\text{ESS,D}}^n}{\eta_{\text{ESS,D}}}\right)\Delta t, \\[8pt]
0 \leq P_{\text{ESS,C}}^n \leq \left(1-u_{\text{ESS}}^n\right) P_{\text{ESS,C}}^{\max}, \\[4pt]
0 \leq P_{\text{ESS,D}}^n \leq u_{\text{ESS}}^n P_{\text{ESS,D}}^{\max}, \\[4pt]
0.2 W_{\text{ESS,R}} \leq W_{\text{ESS}}^n \leq 0.9 W_{\text{ESS,R}}, \\[4pt]
W_{\text{ESS}}^0 = W_{\text{ESS}}^{24}, \\[4pt]
P_{\text{line}}^{\min} \leq P_{\text{grid}}^n \leq P_{\text{line}}^{\max}, \\[4pt]
n = 1,2,\ldots,24.
\end{cases}$$

(9.37)

The cooling, hot water and electric load demand and the output of PV systems can be estimated by the predicted data, and the output of PV can be also calculated by the PV model based on the forecasted solar radiation.

Note that the static optimal scheduling model can be obtained if only one time step is taken into consideration in (9.37). Additionally, by changing the objective function in (9.37), the model can also help to solve other optimization problems with a single objective.

9.2.6 Case Studies

The microgrid shown in Figure 9.1 is taken as an example. The system consists of two micro-turbines (each with 30 kW), a gas boiler of 100 kW,

TABLE 9.1 Parameters of the Key Equipment

Parameter Symbol	Parameter Name	Parameter Value
η_{gen}	Micro-turbine efficiency	0.26
–	Minimum load rate of micro-turbine	30%
α_{gen}	Heat to power ratio	2.8
η_{HR}	Waste heat recovery efficiency	0.75
η_{HX}	Heat exchanger conversion efficiency	0.90
COP_{EC}	Coefficient of performance of electric chiller	4.0
$\eta_{ESS,C}$	Battery charging efficiency	0.95
$\eta_{ESS,D}$	Battery discharging efficiency	0.95
σ_{ESS}	Battery self-discharge rate	0.02%
$W_{ESS,max}$	Upper limit of stored energy of ESS	$0.9W_{ES,R}$
$W_{ESS,min}$	Lower limit of stored energy of ESS	$0.2W_{ES,R}$
$\eta_{GB,steam}$	Gas boiler efficiency	0.90

a PV system of 50 kWp, a battery storage system of 200 kWh, a waste heat recovery boiler of 140 kW, a steam/water heat exchanger of 200 kW and an electric chiller of 300 kW. The related parameters of these devices are shown in Table 9.1.

Figure 9.4 gives the forecasts of the daily demand and the PV output, and Figure 9.5 shows the Time of Use (TOU) electricity price. Meanwhile, assume that the natural gas price is 0.575 \$/m³.

Figures 9.6 through 9.8 give the optimal scheduling results of the demands and supplies of electric, hot water and cooling loads, respectively, and the total operation cost of the microgrid is \$237.82.

Figures 9.6 through 9.8 indicate that the optimal scheduling meets all requirements of the electric, hot water and cooling loads in the example

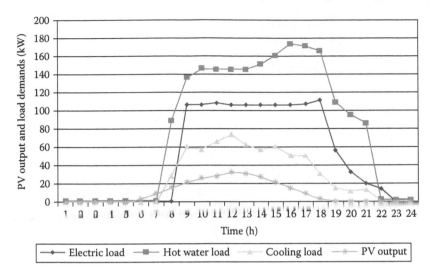

FIGURE 9.4 Forecasts of system daily load demand and PV output.

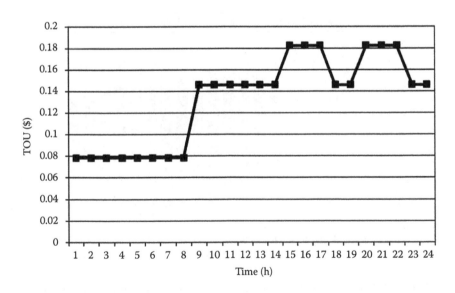

FIGURE 9.5 The TOU electricity price.

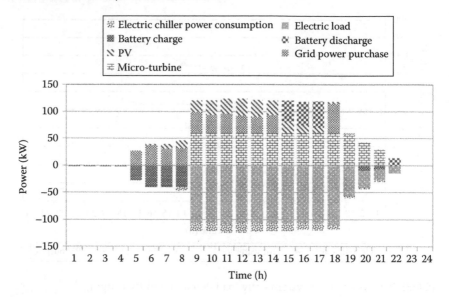

FIGURE 9.6 Optimal scheduling result of the demand and supply of electric power.

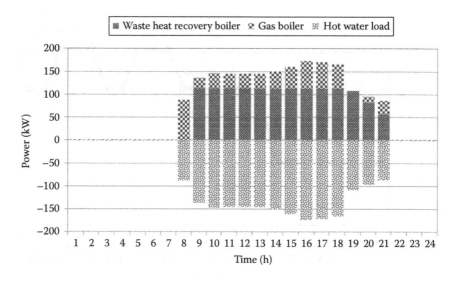

FIGURE 9.7 Optimal scheduling result of the demand and supply of hot water.

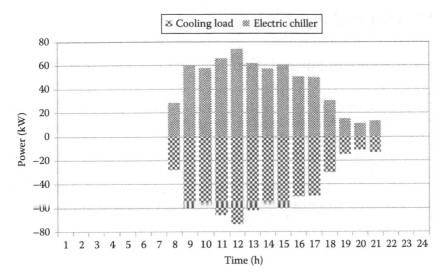

FIGURE 9.8 Optimal scheduling result of the demand and supply of cooling power.

microgrid. The characteristics of the microgrid operation optimization are presented as follows:

1. The battery energy storage system is discharged during the peak price period and charged during the valley price period, which plays the role of peak load shifting. The micro-turbine works only during the plat and peak price periods, and its operation level is mainly limited by the hot water demand, whereas the battery energy storage system can absorb the excess electric energy generated by the micro-turbine.

2. The gas boiler and micro-turbine meet the hot water demand together, while the former only supplements the insufficient heat power of micro-turbine due to the high cost of independent heat supply from the gas boiler.

3. Because the electric chiller is the only cooling source, the cooling load demand is fully supplied by the electric chiller.

9.3 Optimal Planning of Microgrids

9.3.1 Problem Formulation

The optimal planning of microgrids aims to obtain the best solution for the system configuration based on the energy demand and the available distributed energy resources during the planning period together with the specified planning objectives and system constraints. Considering that the dispatch problem can be treated as a sub-problem of the microgrid planning, the optimal planning problem can be compactly expressed as follows:

$$\min F_1(Z),$$
$$\text{s.t.}$$
$$\begin{cases} Z \in \Omega_1, \\ G_1(X,Z) = 0, \\ H_1(X,Z) \leq 0, \\ \begin{cases} \min F_2(X), \\ \text{s.t.} \quad X \in \Omega_2(Z), \\ \qquad G_2(X,Z) = 0, \\ \qquad H_2(X,Z) \leq 0, \end{cases} \end{cases} \qquad (9.38)$$

where
  X is the vector of variables of the dispatch problem
  Z is the vector of variables related to the planning problem, including the types and capacities of devices
  Ω_1 and Ω_2 are the domains of the definition of X and Z
  F_1 and F_2 are the objectives of the planning problem and dispatch problem
  G_1 and H_1 represent the equality and inequality constraints that should be considered in the planning stage
  G_2 and H_2 are the equality and inequality constraints that should be satisfied by the system in each time step of the dispatch period

The constraints G_1 and H_1 of the planning problem include

1. Constraints on the minimum energy utilization efficiency, the maximum amount of carbon emissions and so on

2. Constraints on system economics, such as the total investment cost and the payback period

3. Constraints on available resources, e.g. the allowable installation areas and the capacities of distributed generators like PV and wind turbine generators (WTGs)

4. Constraints on operation performances of the system, such as the power supply reliability, the heating/cooling supply reliability and the energy supply quality

The constraints G_2 and H_2 followed by the dispatch problem mainly contain:

1. Power balance constraints

2. Operation constraints of devices introduced in Sections 9.2.3 and 9.2.4.

As mentioned, the constraints of the dispatch problem should be satisfied throughout the dispatch period. Additionally, the considered operation strategies generally affect the indicators of operation performances of the planning problem discussed, which usually need to be estimated based on the operation results throughout a year or the whole project life cycle. These features make the optimal planning of microgrids more complicated. Here we introduce a two-level approach [11] to the optimal planning problem given in (9.38):

1. *The outer level*: Solve the planning problem.
    This level is intended to optimize the system configuration toward the given objectives, such as the minimum total cost, the maximum energy utilization efficiency, the minimum amount of carbon emissions or their combinations. The optimal solution is determined on the basis of the technical and economic parameters of the system and the devices along with the information generated from the inner level.

2. *The inner level*: Solve the dispatch problem.
    According to the optimal configuration generated from the outer level, this level is designed to determine the optimal dispatch strategy for a typical day with the consideration of the given optimization objective of system operation. Furthermore, the optimal dispatch results, especially the objective value of the operation optimization, are returned to the outer level.

In the two-level approach, the outer level has the optimization horizon of the project life cycle, while the inner usually chooses a number of typical days. Due to the simulation of the actual microgrid operation, the inner level can obtain the operation costs and the operation conditions of the devices throughout a year, such as the reduction of the lifetime of the battery energy storage system (BESS). This operation information contributes to the selection of device types and capacities processed in the outer level.

9.3.2 Objective Function

There are a variety of objectives to be considered for the optimal planning of microgrids, such as indicators of system economics, carbon emissions and energy utilization efficiency [12–14]. Naturally, a single- or multi-objective optimization problem can be built according to the requirements of the stakeholders of the microgrids. A single indicator of system economics is chosen to introduce the optimal planning of microgrids.

The economic objective considered here is to minimize the total investment and operation costs of the microgrids throughout their life cycle. In the following sections, all the costs and revenues are expressed as their present values to take into account the time value of money. In other words, the annual costs and revenues of the project are discounted to the base year at a given discount rate. This approach is convenient for the comparisons of the economics of diverse configurations. Equation 9.39 shows the expression of the objective function which consists of the investment and operation costs of microgrids:

$$f = C_C^{pre} + C_{TEI}^{pre}, \tag{9.39}$$

where
  f is the objective value
  C_C^{pre} and C_{TEI}^{pre} correspond to the present values of the investment and the operation costs

9.3.2.1 Present Value of the Initial Investment Cost of Devices

To calculate the present values, the initial year is selected as the base year for convenience. The total investment cost of devices involves their initial investment costs and replacement costs occurring in the planning period, while the replacement costs need to be discounted to the base year. Consequently, C_C^{pre} can be formulated as follows:

$$C_{C}^{pre} = \sum_{i} \left(C_{I,i} + \sum_{l=1}^{T} \frac{C_{D,i,l}}{(1+r)^{l}} \right), \qquad (9.40)$$

where
$C_{I,i}$ is the initial investment cost of the device i
$C_{D,i,l}$ denotes its replacement cost in year l

If there is no replacement for the device, that term takes the value of 0. r is the discount rate and T is the project life cycle.

9.3.2.2 Present Value of the Operational Cost of Microgrid

The annual microgrid operational cost is composed of the annual cost of purchasing electricity, the annual operation and maintenance (O&M) cost, the annual cost of fuel consumption and the annual revenue of selling electricity, which are denoted as $C_{grid,l}$, $C_{ONM,l}$, $C_{fuel,l}$ and $C_{elec,l}$, respectively. Its present value can be calculated as follows:

$$C_{TEI}^{pre} = \sum_{l=1}^{T} \frac{\left(C_{grid,l} + C_{ONM,l} + C_{fuel,l} - C_{elec,l} \right)}{(1+r)^{l}}. \qquad (9.41)$$

The annual cost of purchasing electricity is related to the local policies of electricity prices. In general, this part consists of three kinds of costs: capacity charge, power charge and electricity charge, as shown in

$$C_{grid,l} = C_{f,l} + C_{d,l} + C_{e,l}, \qquad (9.42)$$

where $C_{f,l}$, $C_{d,l}$ and $C_{e,l}$ refer to the capacity charge, the power charge and the electricity charge in year l.

There may be different regulations of these charges in different countries or regions [15,16]. For example, in China, the capacity charge has been already determined when the users applied for special distribution transformers, and it must be paid monthly in accordance with the applications for the transformer capacities, regardless of whether the electricity is used or not. The power charge is in proportion to the maximum amount of purchasing power from the grid throughout each month, while the electricity charge corresponds to the payment for the total amount of purchasing electricity from the grid. Moreover, the electricity charge is also directly correlated with the electricity prices, which is generally sensitive to the pricing mechanisms in the electricity market.

The annual operation and maintenance cost of the devices can be defined as

$$C_{\text{ONM},l} = \sum_i C_{\text{ONM},i,l}, \tag{9.43}$$

where $C_{\text{ONM},i,l}$ is the operation and maintenance cost of the ith device in year l.

The devices may differ from one to another in the definitions of their O&M costs. For example, the O&M costs of micro-turbines can be split into two parts: fixed cost and variable cost. The O&M costs of PV system and WTGs are normally calculated based on fixed costs per kW, while the corresponding cost of batteries is a function of fixed O&M cost per kWh.

The annual cost of fuel consumption of the devices can be expressed as

$$C_{\text{fuel},l} = \sum_i C_{\text{fuel},i,l}, \tag{9.44}$$

where $C_{\text{fuel},i,l}$ is the annual cost of fuel consumption of the ith device in year l.

Note that the annual costs of purchasing electricity and fuel consumption are calculated on the basis of the optimization results of the dispatch problem. The procedure can be described in two steps: (1) optimize the system operation on typical days in the inner level and then obtain the corresponding costs; (2) estimate the annual costs according to the results of the chosen days at the same level.

9.3.3 Case Studies

An islanded microgrid is taken as a case to introduce the methodology of optimal planning of microgrids. Figures 9.9 and 9.10 present the on-site wind and solar resources and Figure 9.11 illustrates the on-site electricity demand. The structure of the example microgrid is shown in Figure 9.12. In this case, the task of optimal planning is to find an optimal combination of the devices, namely the WTGs, PV system, diesel generating sets (DGSs) and BESS, to meet the demand during the project life cycle.

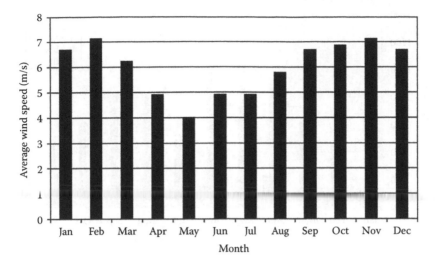

FIGURE 9.9 On-site monthly average wind speed.

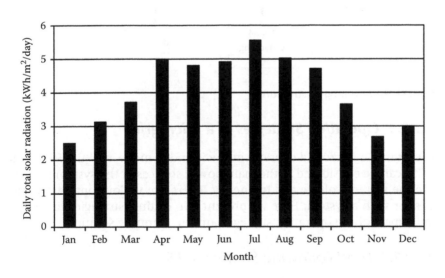

FIGURE 9.10 On-site monthly average daily total solar radiation.

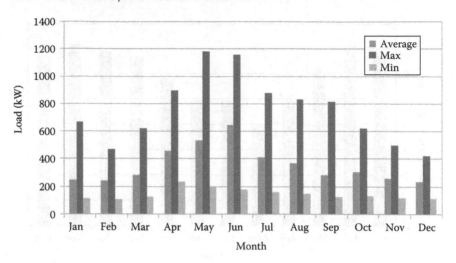

FIGURE 9.11 On-site load demand.

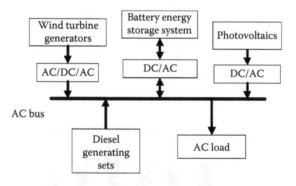

FIGURE 9.12 Microgrid structure.

9.3.3.1 Operation Strategies of a Microgrid

The example microgrid operates in the islanded mode, and the BESS and DGSs can be considered as the main power sources of the system to regulate the system voltage and frequency. There are various optional strategies for such a system. Here we present two distinct strategies to analyse the impact of the operation strategies on the planning solutions.

9.3.3.1.1 Load Following Strategy (LFS)

In this strategy [17,18], the load is first supplied by the WTGs, PV and BESS. The BESS is the main power source and balances the load and renewable energy generation. If the total power generated by the WTGs and PV exceeds the load demand, the batteries will be charged by the excess energy. In contrast, the BESS will be discharged to supply the load when the total renewable energy generation is less than the load demand.

The DGSs act as the main power source in response to variations in the system net load, when the load cannot be met by the WTGs, PV or BESS. Here, 'net load' is defined as the difference between the total renewable energy generation and the load. Considering the power output limits of the DGSs, the DGSs will charge the batteries only if they operate at the minimum power level, and at the same time, the sum of the power output of the DGSs, WTGs and PV is still larger than the load demand. When the load cannot be met by the renewable energy and DGSs even if the DGSs output their allowable maximum power, the BESS will be discharged to supply the load. In the case that the WTGs and PV are capable of meeting the load demand, the DGSs will be shut down and the system will go back into the state where the WTGs and PV together with the BESS are responsible for supplying the load.

9.3.3.1.2 Cycle Charge Strategy (CCS)

The decisions made by this strategy [17,18] are consistent with the LFS's when the renewable energy generation is larger than the load demand, or the WTGs, PV and BESS are able to meet the load demand. If these conditions are not satisfied, the DGSs will be allowed to run and operate in the mode of maximum power output. Under this circumstance, the BESS will be charged whenever possible. In the case that the WTGs, PV and DGSs cannot meet the load demand, the BESS will be discharged to supply the load together with those sources. Note that in this strategy, the DGSs will be shut down when the SOC of the BESS reaches the upper threshold, and then the system will go back into the operating state as mentioned in the LFS.

9.3.3.2 Objective Function of Microgrid Planning

In the example microgrid, the project life cycle is assumed to be 20 years and the minimum net present value of the total system cost is selected as the optimization objective. The objective function is formulated as follows:

$$\min f = C_C^{\mathrm{pre}} + C_{\mathrm{TEI}}^{\mathrm{pre}}. \tag{9.45}$$

The present value of the investment cost of the devices in (9.45) is calculated by

$$C_{\mathrm{C}}^{\mathrm{pre}} = C_{\mathrm{I}} + \sum_{l=1}^{20} \frac{C_{\mathrm{D},l}}{(1+r)^l}, \tag{9.46}$$

where

C_{I} is the sum of the initial purchase costs of the devices, namely the WTGs, PV, BESS, DGSs and converters

$C_{\mathrm{D},l}$ is the sum of the replacement costs of the BESS and DGSs in year l

Considering that the other three types of devices usually have long lifetimes, their replacement costs are ignored.

Since the system is an islanded microgrid, the cost of purchasing electricity is excluded from the system operation cost. The system O&M cost is neglected for the sake of simplicity. Consequently, the system operation cost can be expressed by

$$C_{\mathrm{TEI}}^{\mathrm{pre}} = \sum_{l=1}^{20} \frac{(C_{\mathrm{fuel},l} - C_{\mathrm{elec},l})}{(1+r)^l}. \tag{9.47}$$

9.3.3.3 Results and Discussions

Before we start to optimize the system configuration, it is necessary to preliminarily select some optional types and capacities of the considered devices. Such a strategy can not only reduce the computational complexity of the planning work, but also make the potential solutions more practical. The main considerations in the selection of types and capacities of the devices are as follows:

1. *Diesel generating set*: In the two strategies discussed, the DGSs are used in the case where the WTGs, PV and BESS cannot fulfill the demand. Thus we select the 660 and 1340 kW diesel-generating sets for the following optimal planning of the microgrid in consideration of a peak load of 1183 kW in the example microgrid.

2. *Wind turbine generator*: There are many factors affecting the selection of the WTGs, such as on-site available wind resource, load demand, initial investment cost and maturity of technology. Taking into consideration these factors, we choose two candidate WTGs, the rated capacities of which are 780 and 1500 kW.

3. *Photovoltaics*: The maximum allowable capacity of PV is set as 600 kWp in light of the limitation of on-site available space. Moreover, we

discretize the search range of the optimal installed capacity of the PV with the step of 50 kW.

4. *Battery energy storage*: The lead-acid cells considered in the example microgrid have a rated voltage of 2 V and a rated capacity of 1000 Ah. To satisfy the requirement for the operation voltage of the converters, 300 cells are connected in series with a capacity of 600 kWh. Further, the search ranges of the BESS capacity and the converters capacity are set as 0–5000 kWh and 0–2500 kW. Similar to the consideration of the PV, we also discretize the capacity range of the converters with a step of 250 kW. On the basis of these conditions, the microgrid planning will determine the optimal number of parallel battery strings of the BESS and the optimal capacity of the converters.

Table 9.2 shows the types and economic parameters of the candidate devices. Table 9.3 gives their physical parameters, among which the definitions of the parameters related to the DGSs refer to the model of micro-turbines described in Section 9.2.3. Moreover, the main economic parameters used in the optimal planning are shown in Table 9.4.

Employing this two-level planning method, we present the optimal configurations of the studied microgrid under the LFS and CCS in Table 9.5. For this example, the configurations under the two strategies are basically identical except for the capacity of the BESS. Due to the dissimilarities between the two strategies, the BESS in the configurations

TABLE 9.2 Types and Economic Parameters of Candidate Devices

Types	Initial Investment Cost	Replacement Cost
660 kW DGS ($/unit)	115,000.00	115,000.00
1340 kW DGS ($/unit)	210,000.00	210,000.00
780 kW WTG ($/unit)	1,000,000.00	1,000,000.00
1500 kW WTG ($/unit)	1,620,000.00	1,620,000.00
PV ($/kW)	1,667.00	1,667.00
Lead-acid BESS ($/cell)	317.00	317.00
Converter ($/kW)	450.00	450.00

TABLE 9.3 Physical Parameters of Candidate Devices

660 kW DGS

Lifetime (h)	20,000
Minimum load ratio	0.30
Intercept coefficient of the fuel curve (L/kWh)	0.067
Slope of the fuel curve (L/kWh)	0.179

1340 kW DGS

Lifetime (h)	20,000
Minimum load ratio	0.30
Intercept coefficient of the fuel curve (L/kWh)	0.066
Slope of the fuel curve (L/kWh)	0.191

780 kW WTG

Lifetime (years)	20
Cut-in wind speed (m/s)	4
Rated wind speed (m/s)	13
Cut-out wind speed (m/s)	25

1500 kW WTG

Lifetime (years)	20
Cut-in wind speed (m/s)	3
Rated wind speed (m/s)	11
Cut-out wind speed (m/s)	25

PV

Lifetime (years)	20

(Continued)

TABLE 9.3 (*Continued*) Physical Parameters of Candidate Devices

Lead-Acid BESS

Float lifetime (years)	10
Charge efficiency	0.90
Discharge efficiency	0.90
Minimum SOC	0.40
Maximum SOC	0.90

Converter

Lifetime (years)	20
Efficiency	0.95

TABLE 9.4 Economic Parameters of the Project

Parameters	Values
Project life cycle (years)	20
Discount rate (%)	6.00
Electricity price ($/kWh)	0.33
Diesel price ($/L)	1.32
Cost of control system and project construction ($)	2,430,000

obtained have different capacities and number of replacements during the planning period. In order to quickly transfer the system from the state of DGS running to the state of WTGs, PV and BESS working together, the CCS utilizes the DGSs to charge the BESS when the DGSs operate. This makes the BESS of small capacity more competitive in the optimal planning of the example microgrid. Nevertheless, the CCS unavoidably

TABLE 9.5 Optimal Configurations for the Two Strategies

Devices	Optimal Configurations	
	LFS	CCS
780 kW WTG	0	0
1500 kW WTG	1	1
PV	600 kWp	600 kWp
Lead-acid BESS	3600 kWh (6 strings)	2400 kWh (4 strings)
Converter	1000 kW (4 × 250 kW)	1000 kW (4 × 250 kW)
660 kW DGS	2	2
1340 kW DGS	0	0
Number of replacements of 660 kW DGS	2	2
Number of replacements 1340 kW DGS	0	0
Number of replacements of lead-acid BESS	1	3

shortens the BESS lifetime, which results in a significant increase in the number of replacements of the BESS throughout the planning period. Furthermore, both strategies are inclined to install one 1500 kW WTG, which can not only bring a larger rate of penetration of renewable energy resources, but also reduce the investment cost of the wind generation system. Owing to the limitation of installation space, the capacities of the PV system are 600 kWp in both solutions, which is the value of the maximum available capacity of the PV. For the DGSs, two units, each with a small capacity, are capable of supplying the load more flexibly compared to one unit with a large capacity.

As shown in Figures 9.9 through 9.11, the microgrid has a lower wind resources but heavy load in summer. Consequently, the load cannot

TABLE 9.6 Economic Results of the Optimal Configurations for the Two Strategies

Economics	Values	
	LFS	CCS
Project life cycle (years)	20	20
Initial investment cost ($)	5,370,000.00	5,275,000.00
Present value of replacement cost of the devices ($)	432,923.00	853,454.00
Annual cost of fuel consumption ($/year)	442,413.00	411,260.00
Annual revenue of selling electricity ($/year)	1,030,950.00	1,030,950.00
Internal rate of return (%)	8.13	8.29
Payback period (years)	15.38	14.64

be fully met by the WTGs, PV and BESS most of the time in that period, which means that the DGSs have to operate for a longer time. Moreover, the example microgrid operating under both strategies has to curtail nearly 40% of the annual wind and solar power generation, because the installed capacity of renewable energy sources is approximately twice the peak load demand of the system.

Table 9.6 demonstrates the economic results of the optimal configurations of the microgrid under two strategies. In this case, the system economics under the LFS is slightly worse than the result under the CCS, but the difference between them is insignificant. The solution obtained by the CCS has a larger replacement cost of the devices on account of the increase in the number of replacements of the BESS, as mentioned earlier. However, it should be noted that the DGSs generally operate at high load ratios when they start to run under the CCS, and thus have a high power generation efficiency, which results in the reduction of the amount of annual fuel consumption and therefore the reduction of fuel cost. In general, the configuration under the CCS is superior in terms of the economic indicators of the internal rate of return and the payback period.

☐ Questions

1. List the basic functions of a microgrid energy management system.

2. Explain why dispatch strategies should be considered in microgrid planning and how to model the dispatch and planning together.

3. A simple microgrid is shown in Figure 9.13. The related parameters are listed as follows: Both the hot water load and electric load are 100 kW in the current time period. The ambient temperature is 20°C and the solar radiation is 0.7 kW/m². The rated capacity, derating factor, power temperature coefficient and cell temperature of the photovoltaic system are 50 kWp, 0.9, −0.5%, 50°C, respectively. The rated capacity of both the waste heat recovery boiler and steam/water heat exchanger are 200 kW. The waste heat recovery efficiency and heat exchanger conversion efficiency are 0.75 and 0.90, respectively. The rated capacity, efficiency and heat to power ratio of the micro-turbine are 60 kW, 0.26 and 2.8, respectively.

   a. Calculate the output power of the photovoltaic system.

   b. Calculate the output power and fuel consumption of the micro-turbine for 1 h.

   c. Calculate the electric power purchased from the grid.

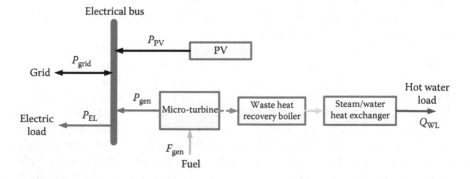

FIGURE 9.13 The microgrid structure diagram for Question 3.

□ References

1. A. Chaouachi, R. M. Kamel, R. Andoulsi et al., Multi-objective intelligent energy management for a microgrid, *IEEE Transactions on Industrial Electronics*, 60(4), 1688–1699, 2013.
2. W. Gu, Z. Wu, R. Bo et al., Modeling, planning and optimal energy management of combined cooling, heating and power microgrid: A review, *International Journal of Electrical Power & Energy Systems*, 54, 26–37, 2014.
3. L. Guo, W. Liu, X. Li et al., Energy management system for stand-alone wind-powered-desalination microgrid, *IEEE Transactions on Smart Grid*, 103(1), 146–154, 2014.
4. H. Ren, W. Zhou, K. Nakagami et al., Multi-objective optimization for the operation of distributed energy systems considering economic and environmental aspects, *Applied Energy*, 87(12), 3642–3651, 2010.
5. C. D. Barley and C. B. Winn, Optimal dispatch strategy in remote hybrid power systems, *Solar Energy*, (4–6), 165–179, 1996.
6. J. Manwell, A. Rogers, G. Hayman et al., Hybrid2 – A hybrid system simulation model: Theory manual, July 2015. Available from: http://www.ecs.umass.edu/mie/labs/rerl/hy2/, accessed on September 11, 2016.
7. R. S. Garcia and D. Weisser, A wind-diesel system with hydrogen storage: Joint optimisation of design and dispatch, *Renewable Energy*, (14), 2296–2320, 2006.
8. T. Lambert and P. Lilienthal, HOMER: The micro-power optimization model. Software developed by NREL, July 2015. Available from: http://homerenergy.com/documents/MicropowerSystemModelingWith HOMER.pdf, accessed on September 11, 2016.
9. E. Mashhour and S. M. Moghaddas-Tafreshi, Integration of distributed energy resources into low voltage grid: A market-based multiperiod optimization model, *Electric Power Systems Research*, (4), 473–480, 2010.
10. V. Kuhn, J. Klemeš and I. Bulatov, MicroCHP: Overview of selected technologies, products and field test results, *Applied Thermal Engineering*, (16), 2039–2048, 2008.
11. C. Wang, L. Guo and P. Li, Optimal dispatch design of microgrid system, in *Handbook of Clean Energy Systems*, Manhattan, John Wiley & Sons, Ltd,Vol. 4, 2015, pp. 1995–2014.
12. L. Guo, W. Liu, J. Cai et al., A two-stage optimal planning and design method for combined cooling, heat and power microgrid system, *Energy Conversion and Management*, 74, 433–445, 2013.
13. L. Xu, X. Ruan, C. Mao et al., An improved optimal sizing method for wind-solar-battery hybrid power system, *IEEE Transactions on Sustainable Energy*, 4(3), 744–785, 2013.
14. R. Dufo-López and J. L. Bernal-Agustín, Multi-objective design of PV-wind-diesel-hydrogen-battery systems, *Renewable Energy*, 33(12), 2559–2572, 2008.
15. J. Wang, C. N. Bloyd, Z. Hu et al., Demand response in China, *Energy*, 35(4), 1592–1597, 2010.

16. A. Picciariello, J. Reneses, P. Frias et al., Distributed generation and distribution pricing: Why do we need new tariff design methodologies, *Electric Power Systems Research*, 119, 370–376, 2015.

17. J. F. Manwell, A. Rogers, G. Hayman et al., *Hybrid2 – A Hybrid System Simulation Model: Theory Manual*. National Renewable Energy Laboratory, Golden, CO, 1998.

18. M. Liu, L. Guo, C. Wang et al., A coordinated operating control strategy for hybrid isolated microgrid including wind power, photovoltaic system, diesel generator, and battery storage, *Automation of Electric Power Systems*, 36(15), 19–24, 2012.

Planning of Smart Distribution Systems

10.1 Introduction

The main objective of traditional distribution system planning is to ensure that the demand is met with an appropriate level of reliability. This is achieved through redundancy in the network so that there is more than one path to supply the demand. A conventional distribution network is operated in open loop with very limited closed-loop control that is applied only to limited aspects of voltage regulation (transformer tap-changers and capacitors). The power flow is one way from the sources to loads. As long as the power demand of loads is well defined, traditional power distribution system planning is relatively straightforward [1].

The introduction of distributed generators (DGs) changes the power flows in a distribution system from one- to two-way flows. Hence, distribution system planning changes from ensuring a simple match between network capacity and load to a coordination of power generation, distribution networks, consumption and even storage. This greatly increases the complexity of the planning process. Distributed generation and loads show different characteristics. If the DG output is intermittent, e.g. from wind or solar generation, then the planning of distribution systems becomes more challenging. Also, an increasing proportion of controllable loads introduces difficulties as consumption is now influenced by factors such as electricity price, demand-side measures and network operating strategies.

In addition to considering the variations in the output of DG and load, the power system planner should also consider the appropriate use of intelligent control devices in order to make sure that the network is operated close to its capacity, thereby reducing the investment cost.

Smart distribution system planning is an optimization process that balances the multiple characteristics of the future power system. Smart distribution systems not only require system planning procedures in response to the connection of new elements (small generators and loads), but also bring many challenges for the detailed planning process.

This chapter begins with a broad comparison between the planning of traditional power distribution systems and smart distribution systems and then introduces the main processes of smart distribution system planning and the detailed techniques involved in the process.

10.2 Planning of Smart Distribution Systems

The planning of distribution systems traditionally involves forecasting electricity demand, locating and sizing substations, designing the power distribution network and assessing of the potential of the grid to supply power. Through electricity demand forecasts, we can determine the total load power, the total energy consumption, the power and energy consumption of each load type, and the spatial distribution of load. The number of substations, their configuration, location and supply area are based on the results of forecasts of the total load and its spatial distribution. The design of the power distribution network includes the structure of the distribution network as well as the selection and layout of circuits (lines, cables and switchgear). It uses the results of load forecasting and substation location and sizing. Assessment of the planned distribution network includes evaluation of reliability and network losses, and compares the advantages and disadvantages of different schemes [2].

When forecasting electricity demand in a smart distribution system, the impact of DGs and controllable load should be considered. The planning of schemes should consider the intermittency of DGs and active control of the demand. This process is shown in Figure 10.1.

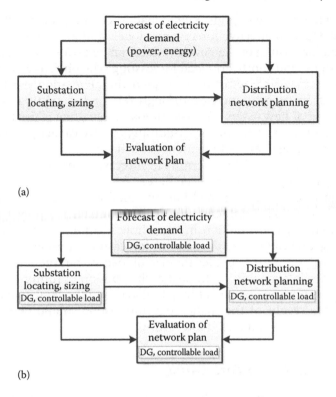

FIGURE 10.1 Planning process of traditional and smart distribution systems: (a) traditional and (b) smart distribution system.

10.3 Load Forecasting of Smart Distribution Systems

10.3.1 Load Forecasting of Traditional Distribution Systems

Load forecasting of the power system is an important part of power system planning [3]. The results of load forecasts provide a reference for regional power development, the construction of electricity infrastructure, energy and resource balancing, and the sharing of power surplus and deficiency between regional power grids. Key to load forecasting is the collection of historical data and effective forecasting models and algorithms.

Load forecasting of the power system can be divided into long-term, medium-term and short-term load forecasting. The objective of medium- and long-term load forecasting is to plan the generation portfolio and power grid, its capacity increase through reinforcing existing assets and the planning of new assets. Short-term load forecasting is used to schedule generator units, determine the spinning reserve capacity and to make reasonable arrangements of maintenance schedules.

The medium- and long-term load forecasting used for distribution network planning [4] is introduced here.

Load forecasting is divided into the total load forecasting and spatial load forecasting. Total load forecasting estimates the power demand and total energy consumption in the whole planning area. Spatial load forecasting, also called load distribution forecasting, considers a particular geographical location, estimates the size of load within the planning area and provides information of the spatial distribution of future load. There is a close relationship between the spatial load and the total load forecasting, and the results of the spatial load forecasting should be consistent with those from the total load forecasting.

10.3.1.1 *Total Load Forecasting*

Methods for total load forecasting include the growth curve model, regression analysis forecasting and time series forecasting.

The time series forecasting method is described in detail here.

The time series forecasting method is based on the historical statistics of load. The load variation with time is used to build a time series model that is then used to estimate the load in the future. A double exponential smoothing time series method is usually used for the load forecasting of distribution networks, because loads can have the characteristic of rapidly increasing over time. The double exponential smoothing method is characterized by low data memory requirement, easy calculation and high forecast precision. The steps of the method are as follows:

Step 1: A single exponential curve fitting is first carried out. It follows the principle of 'preferring the more recent values' to assign different weights to previous observations. This enhances the effect of recent data and weakens the effect of load demand data from long ago. Accordingly, the curve fitting will be more precise at points closer to the present time.

Step 2: Further smoothing is applied to the data after the first exponential smoothing in Step 1, namely modifying the trend of the time series of the smoothed values in order to obtain the double exponential smoothed value.

Step 3: From the smoothed values, two parameters of the double exponential smoothing model are obtained and, hence, the load predicted.

There are several methods described under the name 'double exponential smoothing'. One method referred to as Brown's double exponential smoothing works as follows:

$$S_0' = Y_1$$
$$S_0'' = Y_1$$
$$S_t' = \alpha Y_t + (1-\alpha) S_{t-1}' \tag{10.1}$$
$$S_t'' = \alpha S_t' + (1-\alpha) S_{t-1}''$$
$$(t = 1,2,3,\ldots,T)$$

where

S_t' and S_t'' represent the smoothed values of the single exponential and double exponential at year t

α is the data smoothing factor ($0 < \alpha < 1$)

Y_t represents the historical load data at year t

Normally S_0' is set equal to the historical load data at year 1, i.e. Y_1. S_0'' is also set equal to Y_1.

We calculate two parameters of the smoothing model for the year t according to these two smoothed values, and they are expressed as

$$a_t = 2S_t' - S_t'' \tag{10.2}$$

$$b_t = \frac{\alpha}{1-\alpha} (S_t' - S_t'') \tag{10.3}$$

Then the smoothed value is obtained by

$$F_{t+m} = a_t + mb_t \tag{10.4}$$

where F_{t+m} is the forecast load value at year $t + m$.

Example 10.1

Table 10.1 shows a 10-year historical load data of a certain area. Given a smoothing factor $\alpha = 0.8$, use these data to predict the load data for the next 5 years.

Answer

1. Use Brown's exponential smoothing method to solve this problem. In order to predict the load for 5 years, select $m = 5$.
2. Put the load data into a historical sequence

$$Y = \left[20.0, 21.3, 22.5, 24.0, 25.4, 27.0, 28.6, 30.5, 32.3, 34.4\right]$$

3. Set the initial value $S_0' = S_0'' = Y_1 = 20.0$ and obtain the single exponential smoothing value sequence, S', and the double exponential smoothing value sequence, S'', using Equation 10.1.
4. Calculate the parameters of smoothing model, a and b, using Equations 10.2 and 10.3.
5. The load forecast data for the next 5 years is obtained using Equation 10.4.

$$F_{10+5} = F_{15} = a_{10} + 5b_{10}$$

These calculations are summarized in Table 10.2, and the historical and forecast load are shown in Figure 10.2.

TABLE 10.1 Load Data

Year	2007	2008	2009	2010	2011	2012	2013	2014	2015	2016
Load (MW)	20.0	21.3	22.5	24.0	25.4	27.0	28.6	30.5	32.3	34.4

TABLE 10.2 Load Forecast Calculation

	Year	t	Load (MW) Y	Single Exponential S'	Double Exponential S"	Parameters of Smoothing Model a	b	Forecast m	F
Historical load data	2007	1	20.0	20.0	20.0	20.0	0.00	–	–
	2008	2	21.3	21.0	20.8	21.2	0.83	1	20.0
	2009	3	22.5	22.2	21.9	22.5	1.10	1	22.0
	2010	4	24.0	23.6	23.3	24.0	1.37	1	23.6
	2011	5	25.4	25.0	24.7	25.4	1.40	1	25.4
	2012	6	27.0	26.6	26.2	27.0	1.53	1	26.8
	2013	7	28.6	28.2	27.8	28.6	1.58	1	28.5
	2014	8	30.5	30.0	29.6	30.5	1.79	1	30.2

(Continued)

TABLE 10.2 (*Continued*) Load Forecast Calculation

Year	t	Load (MW) Y	Single Exponential S'	Double Exponential S''	Parameters of Smoothing Model a	b	Forecast m	F
2015	9	32.3	31.8	31.4	32.3	1.80	1	32.3
2016	10	34.4	33.9	33.4	34.4	1.99	1	34.1
2017	11						1	36.4
2018	12						2	38.4
2019	13						3	40.4
2020	14						4	42.4
2021	15						5	44.4

Forecast load

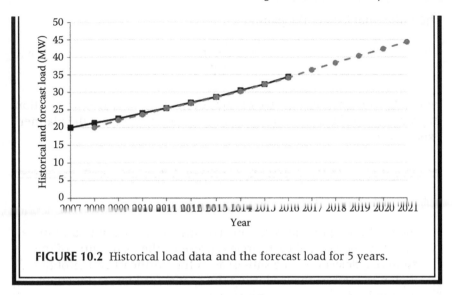

FIGURE 10.2 Historical load data and the forecast load for 5 years.

10.3.1.2 Spatial Load Forecasting

In order to obtain the spatial load forecast, the planning area is divided into different sub-areas according to their land use. The spatial load distribution is found when the load forecast for each individual sub-area is obtained.

The procedure used for this method is as follows:

Step 1: The forecast area is divided into several power supply areas with different land-use types (such as commercial, residential and industrial).

Step 2: By investigating the power consumption of similar sub-areas, a data bank of the load density of different sub-areas is built.

Step 3: Finally, the load density index in each power supply area is determined and the results of spatial load forecast, P_i, are calculated according to Equation 10.5.

$$P_i = \rho_i S_i \quad i \in V \tag{10.5}$$

where
ρ_i is the load density of sub-area i
S_i is the area of sub-area i
V is the area set of the planning area

Then, the total load of the planning area is

$$P = \gamma \sum_{i \in V} P_i \qquad (10.6)$$

where γ is the load diversity factor and is related to the load composition. In practice, γ is obtained from historical data of the area and the planner's experience.

Example 10.2

Figure 10.3 represents a region. The area is divided into a number of sub-areas, and each circle represents the load centre of the sub-area. Select the nine sub-areas in the northeast corner of the region to forecast the load using the spatial load forecast method. Set the load diversity factor γ to 0.7.

Table 10.3 shows the type of loads in each sub-area and its area.

Table 10.4 shows the load density for different types of loads: Calculate the spatial load forecast of the selected sub-areas.

Answer

Calculate the spatial load forecast according to Equation 10.5 The results are presented in Table 10.5.

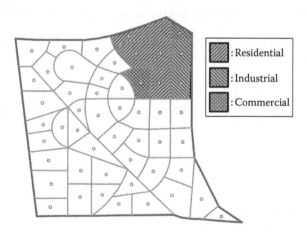

FIGURE 10.3 Regional division of the planning area.

TABLE 10.3 Planning Information of the Sub-Areas in the Planning Area

Sub-area	1	2	3	4	5	6	7	8	9
Load type	R	C	C	R	R	I	C	R	I
Area (km²)	0.91	0.74	0.46	1.04	0.95	0.94	1.01	0.96	0.95

Abbreviations: 'R' stands for 'residential'; 'C' stands for 'commercial'; 'I' stands for 'industrial'. The same abbreviations are used in Table 10.4.

TABLE 10.4 Different Types of Load Density Data

Load Type	I	R	C
Load density (MW/km²)	6	2.5	4

TABLE 10.5 Spatial Load Forecast

Sub-Area	1	2	3	4	5	6	7	8	9
Load (MW)	4.56	3.69	2.74	5.19	4.77	7.54	6.08	4.80	7.58

There are three types of load in the planning area, so the load diversity factor is relatively low. The total load of the planning area can be calculated by Equation 10.6:

$$P = \gamma \sum_{i \in V} P_i = 0.7 \times 46.95 = 32.87 \, (\text{MW})$$

10.3.2 Load Forecasting of Smart Distribution Systems

Load forecasting of smart distribution systems also considers the controllable loads and DGs.

10.3.2.1 Influence of Controllable Load

The *uncontrollable* load cannot be dispatched or respond to the electricity price. The *controllable* load is divided into interruptible and shiftable.

1. *Interruptible load*: This is the load which can be interrupted for a specific period of time. Such loads are controlled through direct load control (DLC) and have a contract with the network operator.

2. *Shiftable load*: This is the load that can be delayed for a period of time. This is achieved through electricity price policies (such as Time of Use pricing and real-time pricing). Electricity users respond to the electricity prices and decide to use their load at a later time. These loads typically include air conditioners, washing machines, electric vehicles, and so on.

3. *Uncontrollable load*: This is traditional load that is connected to the grid when required and cannot be delayed. It includes lighting, farms, public transport and so on. Such a load has existed in the distribution network for a long time.

The relationship between the above three load types is

$$L = L_1 + L_2 + L_3 \tag{10.7}$$

where
  L is the total load
  L_1 is the *interruptible load*
  L_2 is the *shiftable load*
  L_3 is the *uncontrollable load*

The degree of response of shiftable load is closely related to the load demand and electricity price. In general, such load can be divided into *responsive* load and *non-responsive* load, which can be denoted as L_{2A}

and L_{2B}. The ratio of the *responsive* load to the total shiftable load is defined as η, and, hence, the relationship between the two parts of the shiftable load is as follows:

$$\eta = \frac{L_{2A}}{L_2}$$

$$L_2 = L_{2A} + L_{2B}$$

$$L_{2B} = (1-\eta)L_2$$

(10.8)

In a smart distribution system, the ratio of the load that actually responds to the electricity price to the total load is defined as the *controlled load factor* τ_{CL} and it indicates the degree of demand participation. It can be expressed as

$$\tau_{CL} = \frac{L_1 + L_{2A}}{L} \times 100\%$$

(10.9)

The larger the value of τ_{CL}, more load that can be controlled, which means the peak load of the network can be shaved through effective control strategies. With the controlled load factor and load density index, we can forecast the load in a future year.

The load in a smart distribution system can be summarized as shown in Figure 10.4.

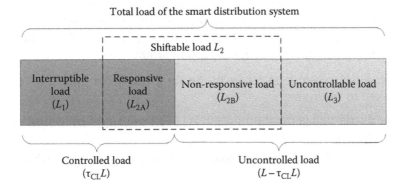

FIGURE 10.4 Load classification of a smart distribution system.

Example 10.3

Select data from sub-area 3 in Example 10.2. The load forecast of this area is 2.74 MW, whereas the value of interruptible load is 0.32 MW, the value of shiftable load is 0.55 MW, and the value of uncontrollable load is 1.87 MW. The proportion of shiftable load that really takes part in response is $\eta = 0.6$. Find the maximum load that the system needs to supply.

Answer

1. From this data, $L = 2.74$, $L_1 = 0.32$, $L_2 = 0.55$ and $L_3 = 1.87$
2. According to Equation 10.8, the responsive part of the shiftable load is

$$L_{2A} = \eta L_2 = 0.33\,\text{MW}$$

3. According to Equation 10.9, the controlled load factor is

$$\tau_{CL} = \frac{L_c}{L} = \frac{L_1 + L_{2A}}{L} \times 100\% = 23.72\%$$

4. Then, the maximum load of this system is

$$L(1 - \tau_{CL}) = 2.09\,\text{MW}$$

10.3.2.2 Influence of Distributed Generation

Distributed generation supports the energy balance in local areas. As distributed generation supplies part of the local load, the power flows in the network are reduced, thus limiting the investment required for network enhancement. The load in the distribution system is divided into the gross system load, net load and distributed generation.

1. The gross system load refers to the total consumer load that is connected to the system.

2. The distributed generation output is the power supplied by distributed generation.

3. The net load is the residual load which is distributed throughout the distribution network.

The net load of users in the smart distribution system will be reduced as part of the system load is supplied by distributed generation. Consequently, when preparing load forecasts, the system load and the net load should be defined clearly.

Example 10.4

Using the data from sub-area 3 in Example 10.2 and the result of Example 10.3, the forecasted gross load is 2.74 MW. We assume that the rated output of photovoltaic power in this area is 1 MW. From the load characteristic of commercial load, the load characteristic curve in a peak load day of sub-area 3 is obtained, which is shown as the dotted line in Figure 10.5. From the statistics and calculation of solar energy data in this area, the photovoltaic output curve is obtained, which is shown as the solid line in Figure 10.5. Obtain the net load.

The difference between the two curves is the load characteristic curve. From Figure 10.5b, the maximum load is 2.1 MW, which occurs at 18:00.

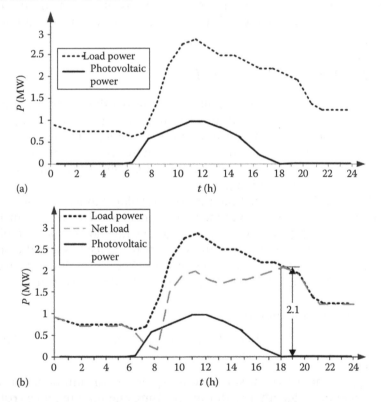

FIGURE 10.5 (a) Load characteristic and photovoltaic output curve of sub-area 3 and (b) maximum load considering DG output.

10.4 Planning of Substations in Smart Distribution Systems

10.4.1 Planning of Substations in Traditional Distribution Systems

The objective of planning substations in traditional distribution systems is to determine the number, geographical location, capacity and power supply area of new substations in the target years. The plan should meet the future load demand and minimize investment and operation cost. However, as the geographical location, power supply area, required capacity and life of new substations are unknown, a number of schemes are possible. In addition, substation planning should also consider the incoming and outgoing feeders and conditions of terrain, traffic, flood control and geology of the substation, which makes planning complicated [5].

10.4.1.1 Mathematical Model

Except locations likely to flood, all other possible sites in an area should be considered. Compared with determining the number and capacity of substations, identifying the location of the substation sites is a more complex matter. Traditional planning of substations involves preparing multiple plans with a number of candidate substation sites and exercising considerable engineering judgement.

The objective of optimal planning of substation is to achieve the optimal comprehensive benefits of investment and operation on the basis of meeting the load demand [6]. Accordingly, the objective function of such a problem should include the investment of the substation and feeders and the operational cost of the substation and the network. The constraints are meeting the load demand and following the principles of secure power supply.

1. The objective function is

$$\min Z = C_{\text{Station}} + C_{\text{Feeder}} + CQ \qquad (10.10)$$

   where

   C_{Station} is the cost of the substations expressed as an annualized investment cost together with the operating and maintenance cost

   C_{Feeder} is the annualized investment in the feeders on the lower voltage side of the substations

If the outgoing feeder is a cable, the investment should include the cable and cable trench; if the feeder is an overhead line, the investment should include the overhead line and overhead equipment; CQ is the estimated annual network losses of the lower-voltage lines of the substation, which is related to the load current and length of the line.

2. The capacity constraints are

$$\sum_{j \in J_i} W_j \leq S_i e(S_i) \cos\varphi \quad i = 1, 2, \ldots, N \qquad (10.11)$$

where
  W_j is the load of node j
  N is the total number of existing and new substations
  S_i is the capacity of the substation i
  $e(S_i)$ is the maximum load factor of the substation i
  $\cos\varphi$ is the power factor of the substation i
  J_i is the set of load nodes supplied by the substation i

If a transformer fails at times of peak load, load on this transformer can be transferred to other remaining transformers or to the lower-voltage circuits fed from an adjacent substation, or consumers can be disconnected. Therefore, the maximum load factor can be over 50%. According to the engineering recommendations in some countries, transformers can be operated in an overloaded condition for a short period of time. For instance, in China, transformers are allowed to operate at a load factor of 130% for 2 h. During this period, the load can be restored or the failure might be repaired. Therefore, for a substation with two main transformers, considering N is 1, the maximum load factor of the main transformers is 65%.

3. Supply radius constraints

$$l_{ij} \leq R_i \qquad (10.12)$$

where
  l_{ij} is the length of the line between substation i and load node j, $l_{ij} = \sqrt{(x_i - x_j)^2 + (y_i - y_j)^2}$, and (x_i, y_i) are the coordinates of substation i and (x_j, y_j) are the coordinates of load node j
  R_i is the power supply radius constraint of substation i

10.4.1.2 Solution Process

The optimal planning of substations considers the number and capacity of the substations, their geographical location and the power supply area. Planning guidelines of distribution systems usually specify the level of redundancy required and which defines the power capacity of the substations and transformers. Consequently, methods such as integer programming are adopted to plan the number and capacity of each substation. The essence of delivering the load in a particular area by the number of substations is to distribute the regional load to the planned substations. As the substation and load are in the same region, methods of spatial geometry such as the Voronoi diagram can be chosen [7,8].

By combining the solution methods, the solution process for the optimal planning of a substation can be obtained. This process is shown in Figure 10.6.

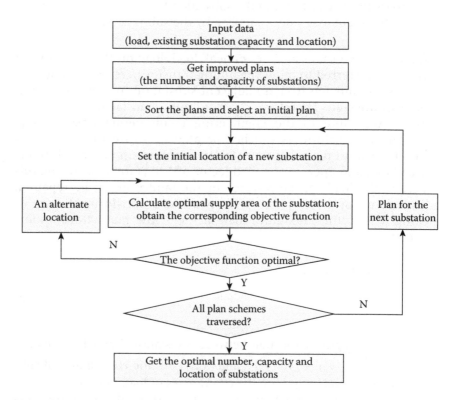

FIGURE 10.6 The typical solution process for the optimal planning of substations in a traditional distribution system.

Example 10.5

As shown in Figure 10.3, the planning area covers 48.85 km² and is divided into 55 sub-areas. The total load of the planning area is projected to be 171 MW in the target year. There is already one 110/10 kV substation in this area whose transformer capacity is 2 × 40 MVA and its location is shown in Figure 10.7. The capacity of the new substations could be chosen from those having 3 × 40, 2 × 40 and 2 × 31.5 MVA transformers.

The cost of constructing a 10 kV overhead line is 27,000 euros/km. A simplified method is used to calculate the total network power losses and the associated cost. It is assumed that the annual cost of network losses when supplying one kilowatt of load is 2.38 euros/kW. Thus the total cost of network losses can be estimated by multiplying the total load and the unit cost €2.38. Table 10.6 presents the cost of substations.

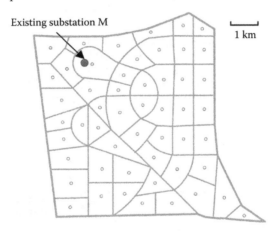

FIGURE 10.7 The division of the planning area and the location of an existing substation.

TABLE 10.6 Cost of Substation Construction

Capacities of Substations (MVA)	3 × 40	2 × 40	2 × 31.5
Cost of substation (million euros)	3.60	2.00	1.70

TABLE 10.7 Substation Capacities of Different Schemes

	3 × 40 MVA	2 × 40 MVA	2 × 31.5 MVA
Scheme 1	1	0	2
Scheme 2	0	3	0

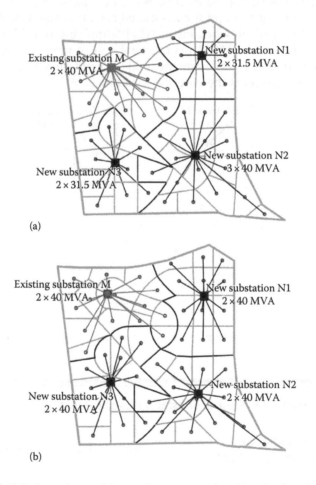

(a)

(b)

FIGURE 10.8 Capacity, position and power supply areas of substations in two plans: (a) scheme 1 and (b) scheme 2.

Obtain the capacity and the locations of the planned substations.

Answer

1. Integer programming is used to determine the new 110 kV substations. The schemes proposed are given in Table 10.7.
2. This method is applied to plan substations in this area, and the planning result is shown in Figure 10.8. The detailed information of load is shown in Table 10.8.
3. Comparing the economics of two planning schemes, the results are shown in Table 10.9.

From the data in Table 10.9, we can find that the total cost of scheme 2 is lower. Thus, scheme 2 is chosen as the final plan.

TABLE 10.8 Substation Load of Different Planning Schemes

		Capacity (MVA)	Load (MW)	Mean Load Factor (Power Factor Is 0.9) (%)
Scheme 1	Substation M	2 × 40	44	61.11
	Substation N1	2 × 31.5	38	60.32
	Substation N2	3 × 40	54	50
	Substation N3	2 × 31.5	35	55.56
Scheme 2	Substation M	2 × 40	38	52.78
	Substation N1	2 × 40	44	61.11
	Substation N2	2 × 40	46	63.89
	Substation N3	2 × 40	43	59.72

TABLE 10.9 Comparison of the Two Schemes

	Substation Cost (Million Euros)	Line Cost (Million Euros)	Network Loss (Million Euros)	Total Cost (Million Euros)
Scheme 1	7.000	1.845	0.407	9.252
Scheme 2	6.000	1.627	0.407	8.034

10.4.2 Planning of Substations in Smart Distribution Systems

Compared with the traditional distribution system, the main changes in the planning of substations in smart distribution systems are taking into consideration the load supplied by the distributed generation and controllable load. Hence, the peak load supplied by the network is reduced and the number, as well as capacity, of substations are also reduced accordingly. Therefore, there is no fundamental change in the objective function and the basic solution process of smart distribution system. Figure 10.9 shows a typical solution process for the optimized substation planning of smart distribution systems.

The darker process box in the flow chart is expanded in the right hand side box. The balance between the substation and load can be determined by load forecast, which was introduced in Section 10.3.

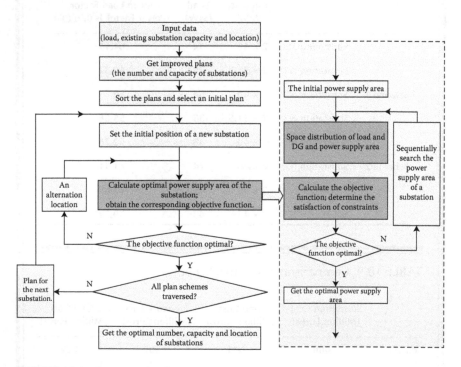

FIGURE 10.9 The typical solution process for the optimized substation planning of smart distribution systems.

Example 10.6

In this example, DGs and controllable load are integrated in the planning area shown in Example 10.5. Their locations are shown in Figure 10.10. Sub-areas marked as A, B and D have some DGs. Sub-area C contains controlled load. Obtain the plan for new substation locations and capacities.

Answer

1. By analysing the load characteristics of each sub-area and their coincidence over time, the load supplied by distributed generation in sub-areas A, B and D, and controllable load in sub-area C, the reduction in the loads can be obtained as shown in Table 10.10.
2. On this basis, the results of the planning of the substation are shown in Figure 10.11 and the detailed information of load is given in Table 10.11.

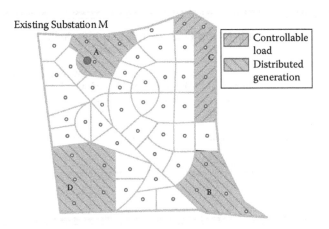

FIGURE 10.10 The planning area.

TABLE 10.10 Load Reduction of the Four Regional Network

Sub-Area Number	A	B	C	D
Load reduction (MW)	5.1	3.2	10.1	7.8

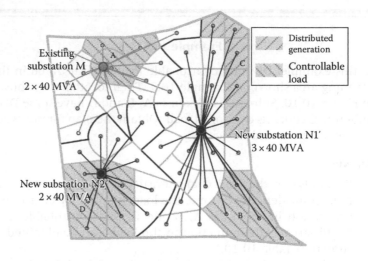

FIGURE 10.11 Capacity, position and the power supply areas of substations considering distributed generation and controllable load.

TABLE 10.11 Substation Load of the Planning Schemes

	Capacity (MVA)	Load (MW)	Mean Load factor (Power Factor is 0.9) (%)
Substation M	2 × 40	34	47.22
Substation N1	3 × 40	64.8	60
Substation N2	2 × 40	46	63.89

TABLE 10.12 Economic Comparison between Active Distribution Network and Passive Distribution Network

	Substation Cost (Million Euros)	Lower Voltage Line Cost (Million Euros)	Cost of Network Losses (Million Euros)	Total Cost (Million Euros)
Example 10.5	6.000	1.627	0.407	8.034
Example 10.6	5.600	1.691	0.407	7.698

> 3. A comparison of the economics of this planning result and the result in Example 10.5 is given in Table 10.12.
>
> From Table 10.12, it can be seen that the peak load is reduced considering the contribution of distributed generation and controllable load. The number and capacity of substations required by the system are reduced. Thus, the investment cost and operation cost are also reduced.

In Examples 10.5 and 10.6, the reliability of the substation is indicated only by the mean load factor. However, if a transformer fails at times of peak load, it would be necessary to reduce the load on the remaining transformer either by transferring load to the lower voltage circuits fed from an adjacent substation or by disconnecting consumers. In many countries, the level of redundancy required in a substation is defined in engineering recommendations which are simple rules based on a statistical assessment of the coincident behaviour of loads (and local generation).

10.5 Network Planning of Smart Distribution Systems

10.5.1 Network Planning of Traditional Distribution Networks

The planning of distribution networks involves the following steps:

Step 1: The network configuration and circuit type is determined.

Step 2: The capacity of the substation and number of outgoing feeders are determined from the total load.

Step 3: The geographic landscape, municipal planning policies and investment and operation costs are considered.

Step 4: Finally, the distribution and direction of outgoing feeders are determined.

10.5.1.1 Selection of Network Configuration and Circuit Type

Each distribution circuit has different configurations depending on the circuit element used, overhead lines or underground cables.

10.5.1.1.1 Overhead Lines

The typical configuration of an overhead line circuit is the multi-segment multi-tie configuration, as shown in Figure 10.12. To meet the $N-1$ reliability criterion, all loads should be supplied through tie-lines in case of a line failure. The feeder maximum load factor of a multi-segment multi-tie configuration is related to the number of tie-lines. It can be calculated using the following equation.

$$\text{Feeder maximum load factor} = 100\% \times \frac{N-1}{N} \tag{10.13}$$

where N is the number of sources of supply.

In Figure 10.12, the feeder has its source busbar and three tie-lines, so the feeder maximum load factor should be set as 75% of its capacity. For instance, if there were a busbar failure, 1/3 of the load could be transferred through each tie-line. After being transferred, the load factor of each tie-line will be 75% + 25% = 100%, and this will not result in overloading.

When tie-line 3 is removed in Figure 10.12, its configuration changes to a multi-segment, two-tie-line one. Similarly, the feeder maximum load factor will be set as $100\% \times (3-1)/3 = 66.7\%$ of its capacity. When tie-lines 2 and 3 are removed in Figure 10.11, its configuration changes to a single-tie mode and the feeder maximum load factor will be set as $100\% \times (2-1)/2 = 50\%$ of its capacity.

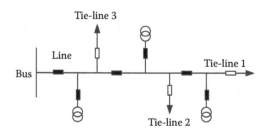

FIGURE 10.12 Multi-segment multi-tie configuration of an overhead line circuit.

The three-segment and two-tie configuration is usually applied in regions with high load density, while the single-tie configuration is usually used in regions with low load density.

10.5.1.1.2 Underground Cables

Numerous configurations are used for cable networks. One of the widely used configurations is the double-ring configuration shown in Figure 10.13. To meet the $N-1$ reliability criterion, in case of line failure, loads should be transferred through the tie-line. The feeder maximum load factor of a double-ring network configuration should be set as 50% of the feeder capacity. For instance, when bus 1 fails in Figure 10.13, all the load of line 1 can be transferred through line 2 by closing the tie switch. The new load factor of line 2 will be 50% + 50% = 100%, and this will not result in overloading.

In the network shown in Figure 10.13, if all the cables and loads connected between bus 3 and bus 4 are removed, the configuration is called a single-ring network configuration and the feeder maximum load factor should still be set at 50% of the feeder capacity.

The double-ring network configuration is normally applied to regions with high load density, while the single-ring network configuration is usually used in regions with low load density.

The selection of configuration and circuit type is determined by the planning principles of the regional power grid, considering factors such as the load density and location of the planning area. In case of low load density in the planning area, in order to control the capital cost, generally the overhead single-tie configuration is chosen. When the load density is high, in order to provide high reliability, the double-ring configuration is chosen.

By comprehensively considering the current-carrying capacity of different types of circuits and the appropriate load factor of different configurations, we can determine the load capacity of each circuit.

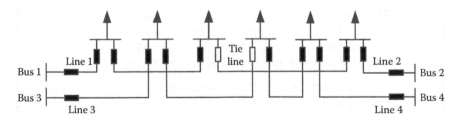

FIGURE 10.13 Double-ring network configuration of underground cables.

Example 10.7

If an overhead line has a current-carrying capacity of 503 A, how much is the carrying capacity of the lines when the configuration of the 10 kV network is (a) multi-segment single tie and (b) multi-segment three tie?

Answer

1. The capacity of the overhead line is $\sqrt{3} \times 10.5\,\text{kV} \times 503\,\text{A} = 9.15$ MVA.
2. On the basis of the $N - 1$ security criterion, the load factors of multi-segment single-tie feeder and multi-segment three-tie feeder are 50% and 75%, respectively. The corresponding carrying capacities of the circuit are approximately 4.57 and 6.86 MVA.

10.5.1.2 Determination of the Number of Outgoing Feeders

After confirming the configuration and circuit type, the ratio between the total load in the forecast year and the circuit load carrying capacity is calculated to determine the total number of outgoing feeders for the planning area. The ratio between the load of the supply area of each substation and the circuit load carrying capacity is used to determine the number of outgoing feeders from each substation.

Example 10.8

In Example 10.5, the following data are given:

> The MV voltage of the planning area is 10 kV.
> The forecast of total load is 171 MW.
> The load that substation M carries is 38 MW.
> The load that substation N1 carries is 44 MW.
> The load that substation N2 carries is 46 MW.
> The load that substation N3 carries is 43 MW.

The configurations that can be selected include the multi-segment single-tie configuration and the multi-segment three-tie configuration. The circuit types that can be selected are overhead line (rated at 503 A) and cable (rated at 461 A).

Obtain the number of outgoing feeders in the whole region and each substation.

Answer

1. Because single tie or single ring networks are selected, according to Example 10.7, the carrying capacity of the circuits are 4.57 MVA (overhead line) and 4.19 MVA (underground cable).

   If the load power factor is 0.9, the maximum active power of the circuit is approximately 4.12 MW (overhead) and 3.77 MW (underground).

2. If the circuits are all overhead lines, the number of regional outgoing feeders is 171/4.12 = 41.5; if the circuits are all underground cables, the number of regional outgoing feeders is 171/3.77 = 45.4. Therefore, there are 42–46 feeders in this region.

3. According to the maximum load each substation can carry, there are 10–11 outgoing feeders of substation M, 11–12 outgoing feeders of substation N1, 12–13 outgoing feeders of substation N2 and 12–13 outgoing feeders of substation N3. The planning results are shown in Figure 10.14.

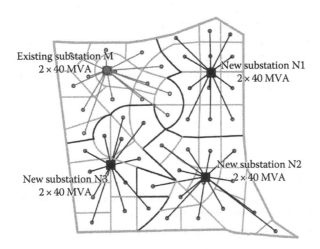

FIGURE 10.14 Substation planning results.

10.5.1.3 Methods for Network Planning

The network planning methods of a distribution system can be divided into manual planning and automatic planning.

The following are the main steps in manual planning:

Step 1: Within the supply area of a substation, determine the circuit directions and obtain several network planning schemes according to the feasibility of circuit corridors that are capable of meeting the power demand.

Step 2: Calculate the economic benefits of different planning schemes, including the investment and operation costs, and select one scheme with the optimal economic benefit.

For the automatic planning of the network, it is essential to take the economic benefits and the basic conditions that should be met as the objective functions and constraints. On this basis, we need to select a network scheme in the feasible solution space, which meets the constraints and achieves the optimal objective.

The objective function of network planning usually refers to the minimal total cost of circuits in the planning year, including the investment and network losses of lines [9,10]. The circuits can also be divided into trunk, tie and branch circuits. The trunk circuit is referred to as the backbone of the network, i.e. main routes from the supply substation down to the load centres, and the branch circuit refers to the service cables that connect between the main routes and the customers. The objective function is as follows:

$$\min f = Z_1 + Z_2 + Z_3 \qquad (10.14)$$

where
  f is the total cost
  Z_1 is the investment and network losses of the trunk circuit
  Z_2 is the investment of the tie circuit
  Z_3 is the investment of the branch circuit [10]

The network losses of tie and branch circuits are not considered, because relatively small currents go through them.

The main constraints of network planning are as follows:

1. No circuit should be overloaded.

2. The voltage at the end of all circuits should meet the requirement for normal operation. For convenience, this requirement is represented by a length limit of trunk circuits. For example, the length of 10 kV circuits should not exceed 5 km.

3. The distribution network should follow the principle that different areas are supplied by different feeders, that is, power supply area of different circuits should not overlap each other.

4. The distribution network should operate as a radial network during a normal operation.

Example 10.9

Take Example 10.5 as an example; the substation planning results are shown in Figure 10.8b, where the capacities of the four substations are all 2 × 40 MVA. The number of outgoing feeders in substation N1, N2, M and N3 is 8, 12, 12 and 12, respectively. The configurations that can be selected include single tie linked and single-ring networks. The construction costs of the overhead line and cable are 27,000 and 130,000 euros/km, respectively. The annual cost of network losses is €2.38 for supplying each kilowatt of load.

Carry out the distribution network planning in this area.

Answer

1. According to the location of the planned substations and circuit corridors in this area, the types of the new 10 kV circuits are shown in Table 10.13, and the layout of the circuits is shown in Figure 10.15.
2. Compare the economics of the two planning schemes above, and the result is shown in Table 10.14.

From the data in Table 10.14, we can find that the total cost of scheme 2 is lower, so scheme 2 is chosen as the final scheme.

TABLE 10.13 Network Planning Schemes to Be Selected

Substation Number	Scheme 1		Scheme 2	
	Circuit Type	Number of Outgoing Feeders	Circuit Type	Number of Outgoing Feeders
Substation M	Overhead line	6	Overhead line	10
	Underground cable	4	Underground cable	0
Substation N1	Overhead line	0	Overhead line	8
	Underground cable	12	Underground cable	4
Substation N2	Overhead line	4	Overhead line	12
	Underground cable	8	Underground cable	0
Substation N3	Overhead line	4	Overhead line	8
	Underground cable	8	Underground cable	4

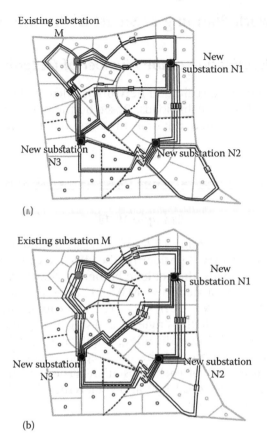

FIGURE 10.15 Circuit layout of two network planning schemes: (a) scheme 1 and (b) scheme 2.

TABLE 10.14 Economic Comparison for Network Planning Schemes

	Investment in Trunk Circuits (Million Euros)	Network Losses (Million Euros)	Investment in Branch Circuits (Million Euros)	Investment in Tie Circuits (Million Euros)	Total Cost (Million Euros)
Scheme 1	20.089	0.407	3.164	1.104	24.764
Scheme 2	8.815	0.407	3.426	0.464	13.112

10.5.2 Network Planning of Smart Distribution Systems

10.5.2.1 Determination of the Number of Outgoing Feeders

The net load of a smart distribution network is calculated recognizing the impact of DGs and controllable loads. Based on the net load, the number of outgoing feeders in a network or a substation is determined.

Example 10.10

Using the data in Example 10.6, the forecast of total load in the target year is 171 MW. Considering the effective capacity of DGs and controllable loads, the net load becomes 144.8 MW. The load that substation M carries is 34 MW, N1' carries 64.8 MW, N2' carries 46 MW and N3' carries 43 MW. The configurations that can be selected include the single tie and single-looped networks. Both the overhead line and underground cables can be considered. Obtain the number of the total regional outgoing feeders and outgoing feeders of each substation. (The current-carrying capacities of the overhead line and underground cable are 503 and 461 A, respectively.)

Answer

1. For a single tie and single-ring network, the maximum active load of the overhead line is 4.12 MW and underground cable is 3.77 MW (see Example 10.8).
2. If all circuits use overhead lines, the total number of outgoing feeders is 144.8/4.12 = 35.1; if all circuits use underground cables, the total number of outgoing feeders is 144.8/3.77 = 38.4. Therefore, the number of regional total outgoing feeders is 36–39.
3. Similarly, substation M has 9–10 outgoing feeders, substation N1' has 16–17 outgoing feeders, and substation N2' has 12–13 outgoing feeders.

10.5.2.2 Methods for Network Planning

Planning of a smart distribution system should also consider load reduction resulting from the introduction of the distributed generation and controllable loads.

Example 10.11

From Example 10.6, after adding the distributed generation and controllable load the resulting substation plan in this region is shown in Figure 10.11.

The number of outgoing feeders of substations M, N1′ and N2′ are 8, 16 and 12, respectively. The configuration that can be selected is single tie or single ring.

1. Based on the location of substations and the line corridors in this planning area, the circuit types that can be selected for 10 kV circuits are shown in Table 10.15 and the layout of the trunk circuits is shown in Figure 10.16.

TABLE 10.15 Network Planning Schemes

	Circuit Type	Number of Outgoing Feeder
Substation M	Overhead line	9
	Underground cable	0
Substation N1′	Overhead line	14
	Underground cable	3
Substation N2′	Overhead line	10
	Underground cable	2

2. The economic comparison between this planning result and the planning result in Example 10.9 is performed, and the results are shown in Table 10.16.

From Table 10.16, we can see that adding distributed generation and controllable loads has reduced the cost by €2.2M.

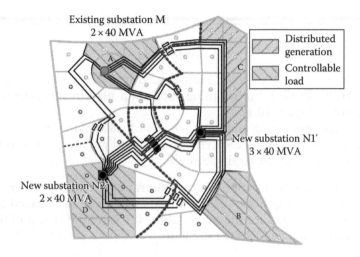

FIGURE 10.16 Network planning schemes considering distributed generation and controllable load.

TABLE 10.16 Economic Comparison for Network Planning Schemes

	Investment in Trunk Circuits (Million Euros)	Network Losses (Million Euros)	Investment in Branch Circuits (Million Euros)	Investment in Tie Circuits (Million Euros)	Total Cost (Million Euros)
Example 10.9	8.815	0.407	3.426	0.464	13.112
Example 10.11	6.556	0.407	3.596	0.345	10.904

10.6 Reliability Assessment of Smart Distribution Systems

10.6.1 Reliability Assessment of Traditional Distribution Systems

A distribution system supplies electric power to all users. It consists of substations connecting different voltage levels, distribution transformers, distribution circuits and other assets.

A distribution system has a great impact on power failures. Statistics show that about 80%–95% of power outages are due to the failure of the distribution system. In modern society, even a power failure of a short duration has a great impact on consumers and society [11]. Therefore, the reliable operation of the distribution system is important. The medium-voltage distribution system plays a more important role in ensuring the reliability of the distribution system than the extra-high-voltage and low-voltage systems.

10.6.1.1 Basic Principle of System Reliability Assessment

A system usually consists of sub-systems or components. These sub-systems or components are connected in series, or in parallel, or as a mixture of both [11].

10.6.1.1.1 Series System

The term 'series system' refers to the functional connection between components that are connected in series and the system can perform its function only if all components perform satisfactorily. As shown in Figure 10.17, all components are involved from the input to the output.

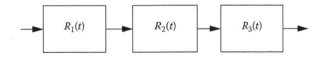

FIGURE 10.17 Three-component series system.

All components in the system have their individual failure rate. The failure rate and reliability of the entire system depend on the components that make up the system. Reliability analysis determines the probability of the system achieving its function over a given time interval. Therefore, the reliability of a series system is the probability that all components have no failure within the given time interval. If the failures of these components are independent, the probability of the system operating successfully (i.e. not failing) is just the product of the probability of each component operating successfully, namely the multiplication rule of probability:

$$R_s(t) = R_1(t) \times R_2(t) \times R_3(t) \times \cdots \times R_n(t) \qquad (10.15)$$

where

$R_s(t)$ is the probability of the successful operation of the system
$R_n(t)$ is the probability of the successful operation of the nth component

10.6.1.1.2 Parallel System

For a 'parallel system', if any component works well, the parallel system can work normally. As is shown in Figure 10.18, it means that there is more than one path from the input to the output.

As the system only requires the normal operation of one of the components that are connected in parallel, other components are redundant. It is such redundancy that leads to high reliability of the system, because for a system failure, all components connected in parallel need to have failed. If the failure of all components is independent, the

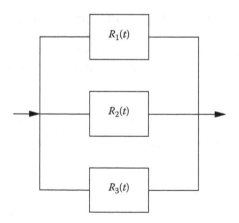

FIGURE 10.18 Parallel connection.

total probability of failure is the product of probability of failure of each component:

$$Q_s(t) = Q_1(t) \times Q_2(t) \times Q_3(t) \times \cdots \times Q_n(t) \qquad (10.16)$$

where
    $Q_s(t)$ is the probability of failure of the system
    $Q_n(t)$ is the probability of failure of the nth component

The reliability of the parallel system is then expressed as

$$R_s(t) = 1 - Q_s(t) \qquad (10.17)$$

Example 10.12

In the series and parallel system shown in Figure 10.19, the reliability parameters of the components are as follows:

$$R(t)_A = R(t)_B = R(t)_C = 0.70; \quad R(t)_D = 0.90; \quad R(t)_E = 0.80$$

Obtain the reliability level of the system.

Answer

System reliability parameter $R_s = R(t)_{A\|B\|C} \times R(t)_D \times R(t)_E$

$$= \left(1 - \left(Q(t)_A \times Q(t)_B \times Q(t)_C\right)\right) \times R(t)_D \times R(t)_E$$

$$Q(t)_A = Q(t)_B = Q(t)_C = 1 - R(t)_A = 1 - 0.7 = 0.3$$

$$R_s = \left(1 - \left(0.3 \times 0.3 \times 0.3\right)\right) \times 0.9 \times 0.8 = 0.701$$

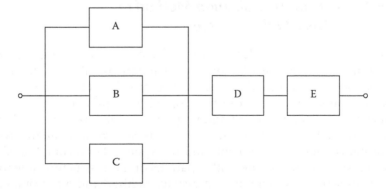

FIGURE 10.19 Series parallel connection of the system in the example.

10.6.1.2 Reliability Indices of Load Points

Distribution system reliability is the ability of the distribution system to provide electric power to users continuously. Users are the final consumers of electricity. Thus, the reliability of a distribution system is calculated according to the power outages of users. The most common outage parameters include the duration of outage U_i and the interruption rate λ_i (failure occurrences/year) within the period of time under investigation (generally a year), and i represents the ith customer.

Common reliability indices of all the load points of a distribution system include system average interruption frequency index ($SAIFI$), system average interruption duration index ($SAIDI$) and average service availability index ($ASAI$) [12]. The calculation formulas of these indices are shown in (10.18) through (10.20):

$$SAIFI = \frac{\text{Total number of customer interruptions}}{\text{Total number of customers served}} = \frac{\sum \lambda_i N_i}{\sum N_i} \quad (10.18)$$

$$SAIDI = \frac{\text{Total time of customer interruption durations}}{\text{Total number of customers served}} = \frac{\sum U_i N_i}{\sum N_i}$$
$$(10.19)$$

$$ASAI = \frac{\text{Customer hours service availability}}{\text{Customer hours service demand}}$$
$$= \frac{8760 \times \sum N_i - \sum U_i N_i}{8760 \times \sum N_i} = 1 - \frac{SAIDI}{8760} \quad (10.20)$$

10.6.1.3 Reliability Evaluation Method of a Distribution System

Methods of reliability evaluation vary depending on the configurations and complexity of the distribution system and the required depth of analysis. For a radial distribution system, the principle of reliability evaluation for the series system can be adopted, i.e. analysing the failure of each component and its effect and preparing the failure mode and effect analysis (FMEA) list [13]. Based on this, it is easy to calculate the overall performance index of load points and the system. For a complicated distribution system, such as one with a parallel or mesh structure, due to the numerous states of system, the state-space method and other simplification methods such as the network simplification and state enumeration are used. The effect of the failure is then analysed and the reliability index of the system can be calculated [14].

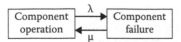

FIGURE 10.20 Outage model of repairable components.

The reliability evaluation of a distribution system involves three aspects: forming a component outage model of distribution system, evaluating the system states and calculating the system reliability [15].

10.6.1.3.1 Outage Model of a Distribution System

A distribution system consists of many components, such as bare overhead conductors, overhead insulated lines, underground cables, circuit breakers, fuses, isolating switches and transformers. Most of them are repairable. The change in the status can be simulated by a two-state circular process of 'operating-outage-operating', as is shown in Figure 10.20. In the figure, λ is the failure rate (e.g. failure occurrences/year) and μ is the repair rate (e.g. number of repairs/hour). Supposing $MTTF$ is mean time to failure (time without failure) and $MTTR$ is the mean time to repair (time with failure), these parameters can be expressed in the following equation:

$$\begin{cases} MTTF = \dfrac{1}{\lambda} \\[2ex] MTTR = \dfrac{1}{\mu} \end{cases} \tag{10.21}$$

Example 10.13

The failure rate of a cable line of a distribution system is 0.065 failures/km·year and the repair rate is 0.2 repairs/h. Obtain the mean time to failure ($MTTF$) and the mean time to repair ($MTTR$) of a 2 km cable.

Answer

$$MTTF = \frac{1}{\lambda} = \frac{1}{0.065 \times 2} = 7.69 \text{ year}$$

$$MTTR = \frac{1}{\mu} = \frac{1}{0.2} = 5 \text{ h}$$

The $MTTF$ of the 2 km cable is 7.69 years, the $MTTR$ is 5 h.

When considering scheduled maintenance in distribution systems, a three-state force outage model, which takes failure and scheduled outages into consideration, can be utilized.

10.6.1.3.2 Evaluation Methods of the System State

The evaluation of the system state follows certain failure guidelines and then analyses the failed state of the system and evaluates the consequence. The most common method of distribution system reliability evaluation is the method of failure mode and effect analysis (FMEA), which is widely applied in reliability engineering. The system states are observed and analysed according to the given reliability criteria and guidelines in order to find the set of failure modes. Then it is necessary to establish the effect of the failure modes, determine the effect of a component failure on the system and thus obtain the system reliability index. The procedure of this method is

Step 1: Select contingencies in the system, i.e. to determine failures of system components that can lead to outages.

Step 2: Power flow analysis and system restoration are performed on each contingency in order to form the failure mode effect table.

Step 3: According to the effects of different failure modes, calculate the reliability index of load points and the system [16].

Example 10.14

A distribution system is shown in Figure 10.21, which consists of three feeders: feeder 1 is a single radial circuit, and feeder 2 and 3 are single tie. Only feeder failures are considered. The number of consumers at each load point is 1. The failure rate of feeders is 0.06 times/km in a year and the mean time to repair is 5 h. Disconnect operation time is 1 h, and restoration is immediate. The length and capacity of each feeder is 1 km and 5.4 MW. The load values are shown in Table 10.17. Derive the failure mode and effect table of this system.

Answer

1. Fault mode effect analysis is used to analyse the effect that a fault on each feeder has on the load nodes. The failure rate of each feeder was obtained by multiplying the unit length failure rate by the length of the feeders. So the fault rate of each of the feeder segments 1–6 is 0.06 times/year.

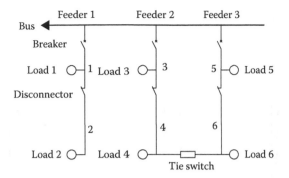

FIGURE 10.21 Example of the calculation of the reliability of a distribu
tion system.

TABLE 10.17 Load Value of Example 10.14

Load Node Number	1	2	3	4	5	6
Load (MW)		2.2	3.2	1.2	2.6	3.9	0.9

2. Take the failure of feeder segment 2 and feeder segment 3, for example, to analyse the consequences of these failures on users.

When there is a failure on feeder segment 2, the circuit breaker of feeder 1 acts, and load 1 and load 2 both will lose power supply. After fault location and isolation, load 1 recovers its power supply. After the fault is repaired, load 2 recovers its power supply. Thus, the outage time of load 1 is the fault isolation time, namely 1 h, and the outage time of load 2 is the mean time to repair of 5 h.

When there is a failure on feeder segment 3, the circuit breaker of feeder 2 operates, and load 3 and load 4 will both lose their power supply. After fault location and isolation, it is necessary to determine whether load 4 can be transferred. The capacity of feeder 3 is 5.4 MW, which is less than the sum of loads 4, 5 and 6, so load 4 cannot be transferred. Therefore, the outage time of loads 3 and 4 is the mean time to repair, namely 5 h.

3. By analysing the failure of the six feeder segments, the failure modes of the power distribution system are obtained, as is shown in Table 10.18.

TABLE 10.18 Table of Failure Mode and Effect Analysis of Example 10.14

Fault Area	Load 1		Load 2		Load 3		Load 4		Load 5		Load 6	
	U_1	λ_1	U_2	λ_2	U_3	λ_3	U_4	λ_4	U_5	λ_5	U_6	λ_6
Feeder segment 1	5	0.06	5	0.06								
Feeder segment 2	1	0.06	5	0.06								
Feeder segment 3					5	0.06	5	0.06				
Feeder segment 4					1	0.06	5	0.06				
Feeder segment 5									5	0.06	1	0.06
Feeder segment 6									1	0.06	5	0.06

The most commonly used distribution system state evaluation methods include minimal cut set [17], logic relation diagram [11] and artificial intelligence [14].

10.6.1.3.3 Distribution System Reliability Index Calculation

After the component outage model and system state evaluation method are determined, the reliability evaluation of the entire distribution system can be carried out.

Example 10.15

For Example 10.14, calculate the system reliability index, SAIFI, SAIDI and ASAI.

Answer

According to the FMEA shown in Table 10.18 of Example 10.14, the influence system components have on the supply received on average by the customers of this network can be intuitively obtained. Using the equations of SAIFI, SAIDI and ASAI, we can directly calculate the reliability index of this network.

$$\lambda_i = \sum_j \lambda_{ij} \ (j \text{ represents feeder segment } j)$$

$$U_i = \sum_j U_{ij} \ (j \text{ represents feeder segment } j)$$

Thus,

$$\lambda_1 = \lambda_2 = \lambda_3 = \lambda_4 = \lambda_5 = \lambda_6 = 0.06 + 0.06 = 0.12 \, (\text{times/year})$$

$$U_1 = U_3 = U_5 = U_6 = 5 + 1 = 6 \, (\text{h/year});$$
$$U_2 = U_4 = 5 + 5 = 10 \, (\text{h/year})$$

$$SAIFI = \frac{\sum \lambda_i N_i}{\sum N_i} = \frac{\begin{array}{c}0.12 \times 1 + 0.12 \times 1 + 0.12 \times 1 + 0.12 \times 1 + 0.12 \\ \times 1 + 0.12 \times 1\end{array}}{1 + 1 + 1 + 1 + 1 + 1}$$

$$= 0.12 \, (\text{times/year})$$

$$SAIDI = \frac{\sum U_i N_i}{\sum N_i} = \frac{6 \times 1 + 10 \times 1 + 6 \times 1 + 10 \times 1 + 6 \times 1 + 6 \times 1}{1 + 1 + 1 + 1 + 1 + 1}$$

$$= 7.33 \, (\text{h/year})$$

$$ASAI = \frac{T \times \sum N_i - \sum U_i N_i}{T \times \sum N_i} = 1 - \frac{SAIDI}{8760} = 1 - \frac{7.33}{8760} = 99.9163\%$$

10.6.2 Effects of New Components in Smart Distribution Systems

10.6.2.1 Effects of Controllable Load

The controllable load of a smart distribution system includes interruptible and shiftable loads. In case of failure in the distribution system, the interruptible and shiftable load that are actually involved in the response can reduce load to be supplied, thus enhancing the reliability level of the system.

Example 10.16

When there are controllable loads in the system shown in Example 10.14, the feeder capacity is 5.4 MW and the load values are as shown in Table 10.19. The proportion of shiftable load which is actually involved in the response is $\eta = 0.6$. When the failure occurs in feeder segment 3, can feeder segment 4 meet the

TABLE 10.19 Load Value in Example 10.16

Load Node Number	1	2	3	4	5	6
Total load (MW)	2.2	3.2	1.2	2.6	3.9	0.9
Interruptible load (MW)	0	0	0	1.4	0	0
Shiftable load (MW)	0	0	0	1.0	0	0
Uncontrollable load (MW)	2.2	3.2	1.2	0.2	3.9	0.9

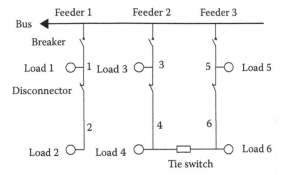

FIGURE 10.22 Distribution system containing controllable load

conditions of transfer? Also, analyse the influence the access of controllable load has on system reliability (Figure 10.22).

Answer

When there is a failure in feeder segment 3, the shiftable load at load node 4 can be delayed. The reduced part of controllable load is

$$L_{2A} = \eta L_2 = 0.6 \times 1.0 = 0.6$$

The controlled load factor is

$$\tau_{CL} = \frac{L_c}{L} = \frac{L_1 + L_{2A}}{L} \times 100\% = \frac{1.4 + 0.6}{2.6} \times 100\% = 76.92\%$$

The maximum load of feeder segment 4 is

$$L(1 - \tau_{CL}) = 2.6 \times (1 - 76.92\%) = 0.6 \, (\text{MW})$$

The remain capability of feeder 3 is

$$C_r = 5.4 - 3.9 - 0.9 = 0.6 \, (\text{MW}) \geq L(1 - \tau_{CL}) = 0.6 \, (\text{MW})$$

Therefore, feeder segment 4 meets the conditions for load transfer.

Analysing the fault of six feeder segments in turn, the table of failure mode and effect analysis can be obtained, which is shown in Table 10.20.

Compared with Table 10.18 in Example 10.4, this example is influenced by controllable load: feeder segment 4 can satisfy the condition for load transfer when the failure of feeder segment 3 occurs. Therefore, the outage time of load node 4 changes from 5 to 1 h without other changes.

TABLE 10.20 Table of Failure Mode and Effect Analysis of Example 10.16

Fault Area	Load 1		Load 2		Load 3		Load 4		Load 5		Load 6	
	U_1'	λ_1'	U_2'	λ_2'	U_3'	λ_3'	U_4'	λ_4'	U_5'	λ_5'	U_6'	λ_6'
Feeder segment 1	5	0.06	5	0.06								
Feeder segment 2	1	0.06	5	0.06								
Feeder segment 3					5	0.06	1	0.06				
Feeder segment 4					1	0.06	5	0.06				
Feeder segment 5									5	0.06	1	0.06
Feeder segment 6									1	0.06	5	0.06

The calculation of the reliability index is as follows:

$$\lambda_i' = \sum_j \lambda_{ij}' \ (j \text{ represents feeder segment } j)$$

$$U_i' = \sum_j U_{ij}' \ (j \text{ represents feeder segment } j)$$

Thus,

$$\lambda_1' = \lambda_2' = \lambda_3' = \lambda_4' = \lambda_5' = \lambda_6' = 0.06 + 0.06 = 0.12 \ (\text{times/year})$$

$$U_1' = U_3' = U_4' = U_5' = U_6' = 5 + 1 = 6 \ (\text{h/year});$$

$$U_2' = 5 + 5 = 10 \ (\text{h/year})$$

$$SAIFI' = \frac{\sum \lambda_i' N_i}{\sum N_i} = \frac{\begin{array}{c} 0.12{\times}1 + 0.12{\times}1 + 0.12{\times}1 + 0.12{\times}1 + 0.12 \\ {\times}1 + 0.12{\times}1 \end{array}}{1+1+1+1+1+1}$$

$$= 0.12 \ (\text{times/year})$$

$$SAIDI' = \frac{\sum U_i' N_i}{\sum N_i} = \frac{6{\times}1 + 10{\times}1 + 6{\times}1 + 6{\times}1 + 6{\times}1 + 6{\times}1}{1+1+1+1+1+1}$$

$$= 6.67 \ (\text{h/year})$$

$$ASAI' = \frac{T \times \sum N_i - \sum U_i' N_i}{T \times \sum N_i} = 1 - \frac{SAIDI'}{8760} = 1 - \frac{7.33}{8760} = 99.9239\%$$

The reliability index comparison of this system and the system in Example 10.14 is shown in Table 10.21.

As we can see, the introduction of controllable loads does not change the system annual average failure frequency (SAIFI), but reduces the system annual average outage time, thus improving the reliability of the system.

TABLE 10.21 Effect of Controllable Load

	With Controllable Load	Without Controllable Load
SAIFI (times/year)	0.12	0.12
SAIDI (h/year)	6.67	7.33
ASAI (%)	99.9239	99.9163

10.6.2.2 Influence of Distributed Generation

Distributed generation can supply local loads, thus reducing the net load and making it easier to transfer load in case of network failure. This can improve the reliability of the system to some extent. Whether load can be transferred is determined by the remaining capacity of the tie feeders, the output of distributed generation and the amount of load to be transferred.

Example 10.17

Using the network of Example 10.14 with the addition of a DG, the distribution system is shown in Figure 10.23. The load curve and the output of the DG connected at feeder segment 4 and the remaining capacity of feeder 3 are shown in Figure 10.24. When there is a failure on feeder segment 3, analyse the load transfer

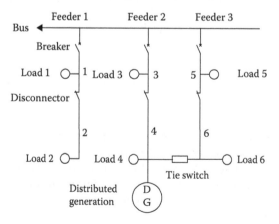

FIGURE 10.23 Distribution system containing distributed generation.

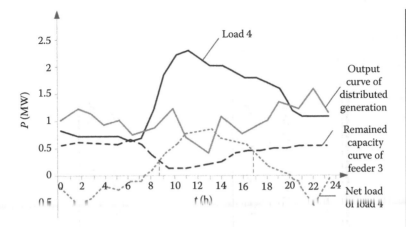

FIGURE 10.24 Load curve and output curve of the DG.

of feeder segment 4 and the influence of access of the distributed generation on system reliability.

Answer

From Figure 10.24, we can see that between 08:20 and 16:20, the sum of the remaining capacity of feeder 3 is less than the net load at the connection point of load 4. Therefore, if feeder segment 3 faults during this period, load 4 cannot be restored. At other times, the remaining capacity of the feeder is greater than the net load. The probability that load 4 cannot be restored is 1/3 (i.e. it is 8 h from 08:20 to 16:20, and this is equivalent to 1/3 of the time of a day), and therefore the expectation of the annual average power failure time of load 4 is

$$T_{EXP} = 5\,h \times \frac{1}{3} + 1\,h \times \left(1 - \frac{1}{3}\right) = 2.33\,h.$$

By comparison, we can find that the network, without distributed generation, meets the condition for load transfer only for a short time (06:06–06:10 a.m.), while during other time periods it is difficult to meet.

Analysing faults on the six feeder segments, the table of failure mode and effect analysis can be obtained, which is shown in Table 10.22.

Compared with Table 10.18, this shows the impact of distributed generation. Feeder segment 4 can meet the condition of load transfer during the failure of feeder segment 3 at certain

TABLE 10.22 Table of Failure Mode and Effect Analysis of Example 10.17

Fault Area	Load 1 U_1''	λ_1''	Load 2 U_2''	λ_2''	Load 3 U_3''	λ_3''	Load 4 U_4''	λ_4''	Load 5 U_5''	λ_5''	Load 6 U_6''	λ_6''
Feeder segment 1	5	0.06	5	0.06								
Feeder segment 2	1	0.06	5	0.06								
Feeder segment 3					5	0.06	2.33	0.06				
Feeder segment 4					1	0.06	5	0.06				
Feeder segment 5									5	0.06	1	0.06
Feeder segment 6									1	0.06	5	0.06

Notes: U_i'' is the annual unavailabilty or outage time (in h/yr) at load point i.
λ_i'' is the failure rate at load point i.

times. Thus, annual average outage time expectancy changes from 5 h to 2.33 h.

The calculation of reliability indices is as follows:

$$\lambda_i'' = \sum_j \lambda_{ij}'' \ (j \text{ represents feeder segment } j)$$

$$U_i'' = \sum_j U_{ij}'' \ (j \text{ represents feeder segmet } j)$$

Thus,

$$\lambda_1'' = \lambda_2'' = \lambda_3'' = \lambda_4'' = \lambda_5'' = \lambda_6'' = 0.06 + 0.06 = 0.12 \ (\text{times/year})$$

$$U_1'' = U_3'' = U_5'' = U_6'' = 5 + 1 = 6 \ (\text{h/year}); \quad U_2'' = 5 + 5 = 10 \ (\text{h/year});$$
$$U_4'' = 2.33 + 5 = 7.33 \ (\text{h/year})$$

$$SAIFI'' = \frac{\sum \lambda_i'' N_i}{\sum N_i} = \frac{\begin{array}{c} 0.12 \times 1 + 0.12 \times 1 + 0.12 \times 1 + 0.12 \times 1 + 0.12 \\ \times 1 + 0.12 \times 1 \end{array}}{1 + 1 + 1 + 1 + 1 + 1}$$

$$= 0.12 \ (\text{times/year})$$

$$SAIDI'' = \frac{\sum U_i'' N_i}{\sum N_i} = \frac{6 \times 1 + 10 \times 1 + 6 \times 1 + 7.33 \times 1 + 6 \times 1 + 6 \times 1}{1 + 1 + 1 + 1 + 1 + 1}$$

$$= 6.89 \ (\text{h/year})$$

$$ASAI'' = \frac{T \times \sum N_i - \sum U_i N_i}{T \times \sum N_i} = 1 - \frac{SAIDI''}{8760} = 1 - \frac{6.89}{8760} = 99.9213\%$$

The comparison of reliability index of this system and Example 10.14 system is shown in Table 10.23.

As we can see, the introduction of distributed generation does not change the system annual average failure frequency, but reduces the system average annual outage time, thus improving the reliability of the system.

TABLE 10.23 Effect of Distributed Generation

	With Distributed Generation	Without Distributed Generation
SAIFI (times/year)	0.12	0.12
SAIDI (h/year)	6.89	7.33
ASAI (%)	99.9213	99.9163

☐ Questions

1. The load forecast results of an area are as follows: total load $L = 2.72$ MW, interruptible load $L_1 = 0.38$ MW, shiftable load $L_2 = 0.65$ MW and uncontrollable load $L_3 = 1.69$ MW. At most 60% of the shiftable load can be shifted to other periods at a time, i.e. $\mu = 0.6$. Calculate the maximum load that the system needs to carry.

2. Describe the optimal planning process of substations in traditional distribution systems and then compare it with that in smart distribution systems.

3. Consider a traditional distribution network with a total load of 146.6 MW. There are, in total, three substations in this network, M, N1′ and N2′, supplying 42, 58.8 and 45.8 MW loads, respectively. Assume that

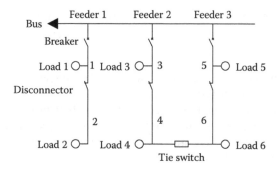

FIGURE 10.25 The distribution system for Question 6.

TABLE 10.24 Load Values for the Distribution System

Load Node Number	1	2	3	4	5	6
Total load (MW)	2.2	3.2	1.2	2.6	0.9	3.9
Interruptible load (MW)	0	0	0	1.4	0	1.6
Shiftable load (MW)	0	0	0	1.0	0	1.0
Uncontrollable load (MW)	2.2	3.2	1.2	0.2	0.9	1.3

the network can be a single tie or single ring network, and the circuit types can be overhead line or underground cable. Calculate the number of outgoing feeders in the entire network and of each substation. (The current-carrying capacity of overhead line and underground cable is 503 and 461 A, respectively. The load power factor is 0.9.)

4. The failure rate of a cable of a distribution system (λ) is 0.075 times/ (km·year) and the repair rate (μ) is 0.2 times/h. Calculate the mean time to failure (MTTF) and the mean time to repair (MTTR) for a 4 km cable.

5. Consider a distribution system containing loads as shown in Figure 10.25. The feeder capacity is 5.4 MW and the load values are listed in Table 10.24. At most 60% shiftable load can be shifted to other periods at a time, that is, $\mu = 0.6$. (1) When a fault occurs in feeder segment 5, does feeder segment 6 meet the conditions of transfer? (2) Calculate the system reliability indices SAIFI, SAIDI and ASAI.

☐ References

1. J. Liang, *Planning Methods of Urban Power Network*. China Electric Power Press, Beijing, China, 2010.
2. C. Niu, *Technology of Load Forecast and Its Application*. China Electric Power Press, Beijing, China, 1998.
3. A. S. Abdelhay and O. P. Malik, *Electric Distribution Systems*, IEEE Press Series on Power Engineering, John Wiley & Sons, 2011.
4. H. He, *Forecast and Management of Electrical Load*. China Electric Power Press, Beijing, China, 2013.
5. Z. Chen and D. Tang, *Planning and Transforming of Urban Electrical Network*. China Electric Power Press, 1998.

6. H. Dai, Y. Yu, C. Huang, C. Wang and S. Ge, Optimal planning of distribution substation locations and sizes – Model and algorithm, *International Journal of Electrical Power and Energy Systems*, 18(6), 353–357, 1996.

7. Y. Lan, *Planning and Construction of Modern Urban Electrical Network*. China Electric Power Press, Beijing, China, 2004.

8. S. Wang, Z. Lu, S. Ge, C. Wang and H. Jia, An improved substation locating and sizing method based on the weighted Voronoi diagram and the transportation model, *Journal of Applied Mathematics*, 81–89, 2014.

9. Ministry of Electric Power Industry Electric Planning and Design Institute, *Electric Power Design Manual*. China Electric Power Press, Beijing, China, 2013.

10. C. R. Bayliss and B. J. Hardy, *Transmission and Distribution Electrical Engineering*, 3rd edn. Boston, MA: Newnes, 2007

11. S. Wang, *Reliability Practical Method and Application of Distribution System*. China Electric Power Press, Beijing, China, 2013.

12. Transmission and Distribution Committee of the IEEE PES, IEEE guide for electric power distribution reliability indices, IEEE Std. 1366, 2001 edn. New York: IEEE Press, 2001.3.

13. R. Billinton and R. Allan, *Reliability Evaluation of Power Systems*, 2nd edn. New York: Plenum Press, 1996.

14. Y. Guo, *Reliability Analysis of Electric System*. Tsinghua University Press, Beijing, China, 2003.

15. R. E. Brown, *Electric Power Distribution Reliability*. New York: Marcel Dekker, 2002.

16. S. Wang and C. Wang, *Modern Distributed System Analysis*. High Education Press, Beijing, China, 2007.

17. R. Billinton and R. Allan, *Reliability Evaluation of Engineering Systems: Concepts and Techniques*, 2nd edn. New York: Plenum Press, 1992.

CHAPTER

DC Distribution Networks

11.1 Introduction

Direct current (DC) has been used for high-voltage transmission circuits since 1954. Power transfer by high-voltage direct current (HVDC) is used to connect asynchronous AC systems and for circuits of underground or submarine high-voltage cables longer than about 50 km (depending on the voltage). HVDC is also considered for the connection of large wind farms far offshore. It becomes an attractive alternative to AC transmission for very long overhead lines with lengths in excess of about 600 km. In 2016, the largest DC link in operation was a 7200 MW overhead line HVDC connection between Jinping and Sunan in China (over 2000 km).

Even though HVDC transmission is now well established, the use of DC at medium (1–30 kV) and low (<~1 kV) voltages is very limited. However, interest in the use of DC for medium- and low-voltage distribution circuits is growing for the following reasons:

- Photovoltaic (PV) panels produce a DC output voltage and before being connected to an AC grid the DC output must be converted to AC. Furthermore, modern wind turbines employ two converters (AC to DC and DC to AC) with a DC bus in between. When connecting renewable energy sources to the existing AC system through multiple power conversion stages, there will be increasing complexity and reduced efficiency. Instead, if the renewable energy sources are directly connected to DC networks, the power loss in converters can be eliminated, thus saving 2.5%–10% of the energy transferred [1].

- Many distributed generation systems, microgrids and energy storage devices use power electronic interfaces to the AC grid. These sources are compatible with DC distribution systems where the energy can be used directly to supply loads and/or energy storage devices. This results in reduced conversion losses [2].

- To provide reliability against unplanned AC outrages, uninterruptible power supplies (UPS), which are essentially battery storage systems with a DC bus, are employed. Furthermore, electric vehicles (EVs) are not only significant consumers of electrical power, but their batteries could also provide improved grid reliability and energy storage whenever they are grid connected [3]. The number of conversion stages associated with these distributed energy storage devices is reduced if they are used with a DC network.

- With the unprecedented development of electronic technology, DC-powered consumer electronic devices, such as computers, televisions and cordless tools, have become a significant part of the system load. As the energy consumption of consumer electronic equipment increases, supplying energy through the traditional AC system will become less efficient and more complex, because more of the energy must be converted to DC.

- Overall energy efficiency can be enhanced when in addition to a low-voltage AC network, buildings and homes also have a low-voltage DC supply. In the future, homes will have heat-pumps, photovoltaic systems and charging points for EVs. All these devices use power electronic converters with filters to connect to the AC grid. Major cost savings can be realized by avoiding the complex converter systems.

11.2 DC Transmission

DC transmission can be used to interface two AC networks or to bring power from a DC source such as a PV plant to load centres.

The majority of HVDC transmission systems installed to date use line-commutated, current source converters (LCC-CSC). Thyristors are used as switching devices within LCC-CSC and can handle high currents, are reliable in operation and have relatively low on-state losses. In LCC-CSC a constant direction of DC is maintained and its voltage is controlled by the switching of the thyristors, thus controlling the amount and direction of the power transferred [3].

Over the past few decades, advances in power semiconductor devices have led to self-commutated static power converters that are becoming commercially available. These power converters, often referred to as voltage source converters (VSCs) most commonly use insulated-gate-bipolar transistors (IGBTs) as the switching devices. In a VSC system, a constant direct voltage is maintained and the current is controlled by the switching of the devices, thus controlling power transfer.

For medium-voltage DC (MVDC) applications, it is anticipated that VSCs will be the dominant technology. Figure 11.1 shows a typical MVDC connection. As can be seen, two VSCs are connected through a DC circuit to the AC systems.

As in AC circuits, a variation of the load power (P_L) influences the voltage of the MVDC lines (V_{DC}). Therefore, one simple control strategy of the converters is to maintain a constant DC voltage at one end and deliver the required power at the other end. Usually, two control loops (outer control and inner control), as shown in Figure 11.2, are employed to achieve this control strategy.

Vector control schemes are usually employed for the inner current control [4]. A conventional vector control scheme uses direct and quadrature (dq) components that are constant in the steady state rather than three-phase sinusoidal signals. This allows the use of simple controllers often using proportional and integral (PI) gains. A phase-locked loop (PLL) is used to lock the dq quantities onto the phases of the AC system. Often, the d-axis is chosen as to coincide with the phase A voltage, thus $V_{sq} = 0$. This controller with a pulse width modulated (PWM) firing pulse generator is shown in Figure 11.3.

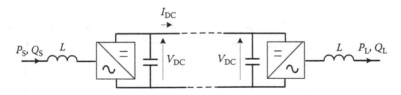

FIGURE 11.1 MVDC scheme.

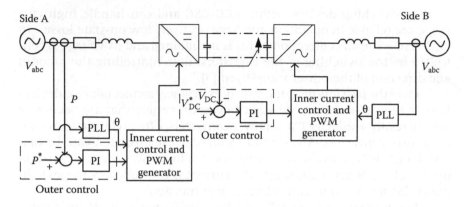

FIGURE 11.2 Conventional controller for AC-DC converter.

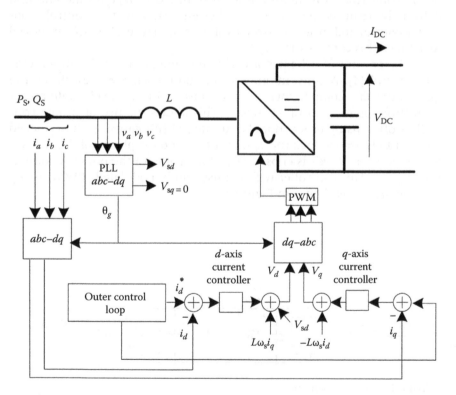

FIGURE 11.3 Conventional d–q controller for inner control.

Example 11.1: Power Flow in a DC Circuit

One side of an MVDC scheme is shown in Figure 11.4. The VSC is connected to a three-phase network having a voltage of 20 kV (L–L). Active and reactive power transfer to the network are 5 MW and 3 Mvar, respectively. Calculate the voltage that should be generated by the VSC. The value of the inductor (L) is 100 mH. Neglect all losses.

Solution

$$\text{With } S_{base} = 5 \, MVA \quad \text{and} \quad V_{base} = 20 \text{ kV,}$$

$$Z_{base} = 20^2/5 = 80 \, \Omega$$

$$P = 5/5 = 1 \text{ pu}$$

$$Q = 3/5 = 0.6 \text{ pu}$$

$$\mathbf{S} = P + jQ = 1.0 + j0.6 \text{ pu}$$

$$\mathbf{X} = j\frac{100\pi \times 100 \times 10^{-3}}{80} = j0.393 \text{ pu}$$

Since $\mathbf{S} = \mathbf{VI}^*$, $\mathbf{I}^* = 1.0 + j0.6$
    Therefore, $\mathbf{I} = 1.0 - j0.6$
    The converter voltage is given by

$$\mathbf{V_C} = 1 + \mathbf{IX} = 1 + (1.0 - j0.6)(j0.393) = 1.2358 + j0.393 \text{ pu}$$

$$\mathbf{V_C} = 25.91 \angle 17.5° \text{kV}$$

Alternatively, the converter voltage can be calculated using the following equations for the active and reactive power.

$$P = \frac{V_S V_C}{X} \sin \delta$$

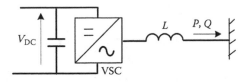

FIGURE 11.4 Network for Example 11.1.

$$Q = \frac{V_S \left(V_C \cos\delta - V_S \right)}{X}$$

where

V_S is the network source voltage

V_C is the converter voltage

δ is the converter voltage angle

Substituting values, we get

$$P = \frac{\left(20 \times 10^3 / \sqrt{3} \right) V_{C1}}{100\pi \times 100 \times 10^{-3}} \sin\delta = 367.55 V_{C1} \sin\delta = \frac{5 \times 10^6}{3}$$

$$V_{C1} \sin\delta = 4534.5$$

$$Q = \frac{\left(20 \times 10^3 / \sqrt{3} \right) \left(V_{C1} \sin\delta - 20 \times 10^3 / \sqrt{3} \right)}{100\pi \times 100 \times 10^{-3}} = \frac{3 \times 10^6}{3}$$

$$V_{C1} \cos\delta = 14267.7$$

Then

$$V_{C1} = \sqrt{3} \times \sqrt{4534.5^2 + 14267.7^2} = 25.9 \, \text{kV}$$

$$\delta = \tan^{-1}(4534.5 / 14267.7) = 17.6°$$

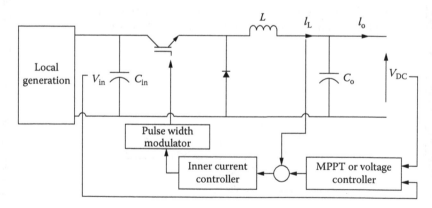

FIGURE 11.5 DC–DC converter for MVDC distribution.

11.3 Connection of Renewable Generating Plant

In order to transmit power from a PV plant that generates DC, a DC–DC converter is employed at the source end. As the power flow is always from the source to the network, a unidirectional DC–DC converter, as shown in Figure 11.5, can be used.

Typically the DC–DC converter uses an inner current loop which generates the pulse width modulation signals. The outer-loop controllers differ from one energy source to another but for a PV system will use a maximum power point tracker (MPPT).

Example 11.2: Power Flow in a DC Circuit

The PV array shown in Figure 11.6 produces 500 kW. If $V_{d2} = 1\,\text{kV}$, calculate the voltage that should be maintained at V_{d1} to ensure power transfer. Also, calculate the current through the line. The resistance of the circuit is 0.2 Ω.

Solution

The power transfer in the line is given by

$$P = V_{d1}\left[\frac{V_{d1} - V_{d2}}{R}\right]$$

$$V_{d1}^2 - V_{d2}V_{d1} - RP = 0$$

$$V_{d1} = \frac{V_{d2} \pm \sqrt{V_{d2}^2 + 4RP}}{2}$$

$$\text{Since } V_{d1} > V_{d2}, \quad V_{d1} = \frac{V_{d2} + \sqrt{V_{d2}^2 + 4RP}}{2}$$

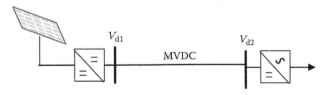

FIGURE 11.6 PV connection through an MVDC line.

Substituting value, we get

$$V_{d1} = \frac{V_{d2} + \sqrt{V_{d2}^2 + 4RP}}{2}$$

$$= \frac{1 \times 10^3 + \sqrt{1 \times 10^6 + 4 \times 0.2 \times 500 \times 10^3}}{2}$$

$$= 1.092 \, kV$$

Since $P = V_{d1}I_{d1}$, $I_{d1} = 500/1.092 = 458$ A

11.4 Faults on VSC-Based MVDC

11.4.1 AC-Side Faults

A VSC in an MVDC scheme can be arranged to produce only 1 pu fault current during an AC-side fault. This is the cheapest arrangement and protects the converter from damage. However, if a larger fault current is required, then higher-rated IGBTs should be used in the converter.

11.4.2 DC-Side Faults

In the MVDC scheme shown in Figure 11.1, consider a fault on the DC circuit. If no protection is employed on the DC side then the circuit shown in Figure 11.7 exists under fault conditions [5]. Even though the inductance of the cable is not effective under steady DC flow, it will have an impact on the transient current during a fault.

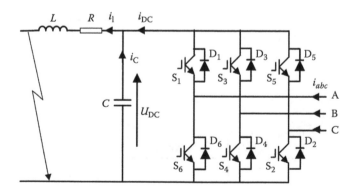

FIGURE 11.7 A circuit of a VSC DC link with DC circuit fault.

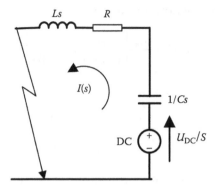

FIGURE 11.8 Equivalent circuit immediately after the fault (ignoring the contribution from the VSC bridge).

Immediately after the fault, the capacitor, C, will discharge into the fault. The fault current will increase and the DC-side voltage will decrease. The initial few milliseconds after the fault are analysed by transforming the RLC circuit into the Laplace domain, as shown in Figure 11.8. In this circuit, the contribution from the VSC is neglected, thus $i_1 = i_C = I$. It was assumed that just before the short circuit, the capacitor is fully charged to a voltage U_{DC}.

Applying Kirchhoff's voltage law,

$$LsI(s) + RI(s) + \frac{I(s)}{Cs} - \frac{U_{DC}}{s} = 0$$

$$I(s) = \frac{CU_{DC}}{LCs^2 + RCs + 1} = \frac{U_{DC}}{L}\left[\frac{1}{s^2 + (R/L)s + (1/LC)}\right] \qquad (11.1)$$

The denominator of Equation 11.1 is written in the following form:

$$s^2 + \frac{R}{L}s + \frac{1}{LC} = (s+\alpha)(s+\beta)$$

where

$$\alpha = \frac{1}{2}\left[\frac{R}{L} - \sqrt{\frac{R^2}{L^2} - \frac{4}{LC}}\right]$$

$$\beta = \frac{1}{2}\left[\frac{R}{L} + \sqrt{\frac{R^2}{L^2} - \frac{4}{LC}}\right]$$

Taking partial fractions of Equation 11.1,

$$I(s) = \frac{U_{DC}}{L}\left[\frac{A}{(s+\alpha)} + \frac{B}{(s+\beta)}\right] \tag{11.2}$$

where

$$A = -B = \frac{1}{(\beta - \alpha)} = \frac{1}{\sqrt{(R^2/L^2) - (4/LC)}}$$

Taking the inverse Laplace transform of Equation 11.2

$$i(t) = \frac{U_{DC}}{\sqrt{R^2 - 4L/C}}\left[e^{-\alpha t} - e^{-\beta t}\right] \tag{11.3}$$

The shape of $i(t)$ depends on α and β.

When $R^2 > 4L/C$, α and β are real and therefore $i(t)$ changes exponentially. When $R^2 < 4L/C$, α and β are complex. Then

$$i(t) = \frac{U_{DC}}{\sqrt{4(L/C) - R^2}} e^{-(R/2L)t} \sin\left[\left(\frac{1}{2}\sqrt{\frac{4}{LC} - \frac{R^2}{L^2}}\right)t\right]$$

In the Laplace domain, the capacitor voltage is given by

$$U_{DC}(s) = \frac{AU_{DC}}{L}\left[\frac{1}{(s+\alpha)} - \frac{1}{(s+\beta)}\right][R + Ls]$$

$$= \frac{U_{DC}}{L}\left[\frac{AR - AL\alpha}{(s+\alpha)} - \frac{AR - AL\beta}{(s+\beta)}\right]$$

Taking the inverse Laplace transform

$$U_{DC}(t) = \frac{AU_{DC}}{L}\left[(R - L\alpha)e^{-\alpha t} - (R - L\beta)e^{-\beta t}\right] \tag{11.4}$$

From Equation 11.4, it can be seen that when $t = 0$

$$U_{DC}(t) = \frac{AU_{DC}}{L}[R - L\alpha - R + L\beta]$$

$$= AU_{DC}[\beta - \alpha]$$

Since $A = \frac{1}{(\beta - \alpha)}$, $U_{DC}(t) = U_{DC}$

Example 11.3

An MVDC scheme uses a DC capacitor of 1000 μF and a XLPE cable of 185 mm². The cable has the following parameters:

$$\text{Resistance per km} = 0.15\,\Omega$$
$$\text{Inductance per km} = 0.05\,\text{mH}$$

If a fault occurs 10 km from the VSC, obtain expressions for the fault current and DC link voltage immediately after the fault. Hence, draw the wave forms of current and voltage up to 10 ms. Assume that before the fault the DC link voltage is 10 kV.

Solution

$$\alpha = \frac{1}{2}\left[\frac{R}{L} - \sqrt{\frac{R^2}{L^2} - \frac{4}{LC}}\right]$$

$$= 0.5\left[\frac{0.15 \times 10}{0.05 \times 10^{-3} \times 10}\right.$$

$$\left. - \sqrt{\left(\frac{0.15 \times 10}{0.05 \times 10^{-3} \times 10}\right)^2 - \frac{4}{0.05 \times 10^{-3} \times 10 \times 1000 \times 10^{-6}}}\right]$$

$$= 1000$$

$$\beta = \frac{1}{2}\left[\frac{R}{L} + \sqrt{\frac{R^2}{L^2} - \frac{4}{LC}}\right]$$

$$= 0.5(3000 + 1000) = 2000$$

$$A = \frac{1}{(\beta - \alpha)} = \frac{1}{2000 - 1000} = 1 \times 10^{-3}$$

$$i(t) = \frac{1 \times 10^{-3} \times 10 \times 10^3}{0.05 \times 10^{-3} \times 10}\left[e^{-1000t} - e^{-2000t}\right]$$

$$= 20\left[e^{-1000t} - e^{-2000t}\right]\text{kA}$$

$$U_{DC}(t) = \frac{1 \times 10^{-3} \times 10 \times 10^3}{0.05 \times 10^{-3} \times 10}\left[(0.15 \times 10 - 0.5)e^{-\alpha t} - (0.15 \times 10 - 1)e^{-\beta t}\right]$$

$$= 20\left[e^{-1000t} - 0.5e^{-2000t}\right]$$

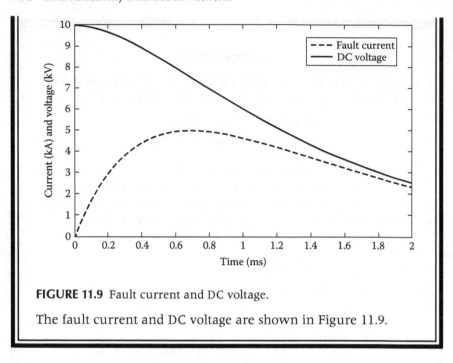

FIGURE 11.9 Fault current and DC voltage.

The fault current and DC voltage are shown in Figure 11.9.

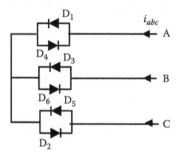

FIGURE 11.10 An equivalent circuit when the DC capacitor is fully discharged.

As the DC voltage drops, the freewheeling diodes, D_1–D_6, become forward biased. Therefore, in addition to the capacitor discharge current, another current passes from the AC side to the DC side through the freewheeling diodes. When the capacitor is fully discharged, it acts as a short circuit to the AC side through the freewheeling diodes, as shown in Figure 11.10. This results in a large current, similar to a three-phase fault current, flowing on the AC side.

Example 11.4

For the case given in Example 11.3, sketch the phase and DC voltages when the diodes D_1, D_3 and D_5 start conducting after a DC fault. Assume that the fault occurs at the positive zero crossing of the Phase A voltage.

Solution

Figure 11.11 shows the three phase voltages and DC voltage. When Phase A voltage exceeds the DC, diode D_1 starts conducting. When the DC capacitor is fully discharged, any forward biased diode in Figure 11.10 will conduct.

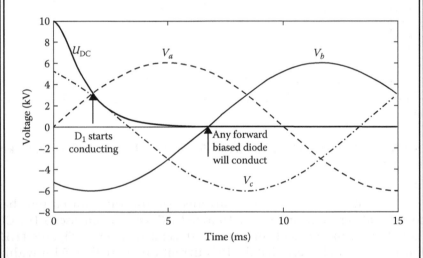

FIGURE 11.11 Voltage after a fault.

The system described in Example 11.3 was simulated in EMTDC/PSCAD software. Figures 11.12 and 11.13 show the DC-side voltage and fault current when the fault is 5 and 10 km from the VSC. The fault is applied 8 s after the start of the simulation.

As fault current on the DC side increases just after a fault, overcurrent protection can be employed on the DC side if a suitable DC circuit breaker (CB) is available.

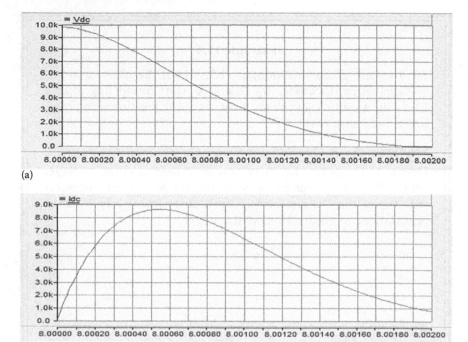

(a)

(b)

FIGURE 11.12 (a) DC voltage and (b) fault current when the fault occurs 5 km from the VSC.

It is difficult to break DC currents because the fault current has no natural zero-crossing point. To avoid DC system collapse, a DC CB should interrupt the fault current as fast as possible. From Figures 11.12 and 11.13, it can be seen that the DC current can more than 5 kA within 200–700 μs and the voltage can collapse within 2 ms. Therefore, the operating speed of the CB used should be less than a few milliseconds. The solid-state CB shown in Figure 11.14 operates quickly and thus is suitable for MV applications. However, a number of series-connected IGBTs are required to withstand the system voltage.

11.5 Multi-Terminal DC Grids

A multi-terminal DC scheme uses a common DC network to which AC systems are connected through VSC and DC renewable energy plant through DC–DC converters. A scheme with four terminals is shown in Figure 11.15.

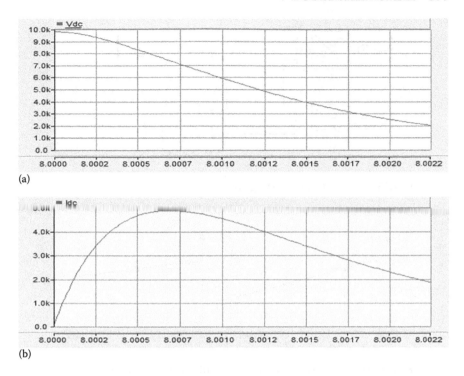

(a)

(b)

FIGURE 11.13 (a) DC voltage and (b) fault current when the fault occurs 10 km from the VSC.

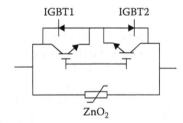

FIGURE 11.14 Solid-state circuit breaker.

The usual principle of operation is that one converter controls the DC voltage while the others regulate their power transfers. However, in order to set the power references, communication between the converters may be required. Another method is to use a voltage droop control [6]. VSC1 is controlled to maintain a constant DC voltage. The DC–DC converter operates on MPPT to extract maximum power from the PV plant. VSC2 and VSC3 operate on droop to set the power transfer, as shown in Figure 11.16.

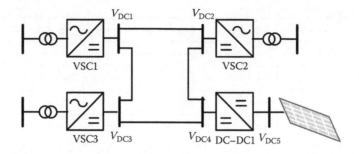

FIGURE 11.15 A multi-terminal DC grid.

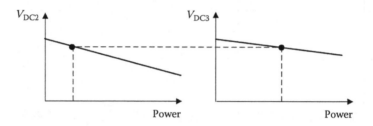

FIGURE 11.16 Droop controller of a four-terminal MVDC circuit.

11.6 DC Microgrids

A DC microgrid is a small power system consisting of local generation and loads with several DC feeders. Such a system is coupled to the utility network at a point through a VSC. A DC microgrid can operate either autonomously or in grid-connected mode.

A number of energy storage units and local generation units such as photovoltaic, wind turbines and small-scale thermal generation can be used in a DC microgrid. The capacity of the local generation would typically be less than 500 kW. In order to interface these sources, to the microgrid, either DC–DC or AC–DC converters are used.

Demand–supply matching is achieved easily in grid-connected mode. However, it becomes difficult during autonomous operation. Transients need to be managed. For these two reasons, energy storage is essential in a DC microgrid. Energy storage is usually a combination of high power and high energy storage. A high energy storage device, such as a battery, is used to overcome medium- and long-term generation mismatches. On the other hand, high power storage devices, such as supercapacitors, are used to manage short-term transients.

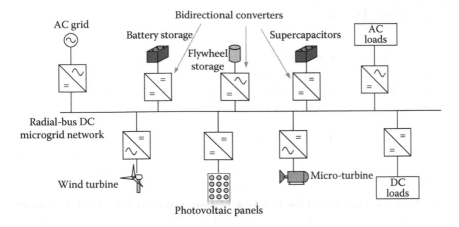

FIGURE 11.17 Radial-bus DC microgrid.

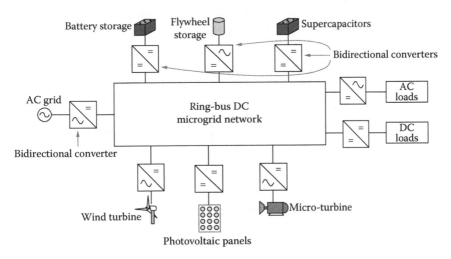

FIGURE 11.18 Ring-bus DC microgrid.

A network of a DC microgrid can have two forms, as shown in Figures 11.17 and 11.18. A radial feeder network is used in Figure 11.17. Often, several feeders are used in a radial network.

Figure 11.18 shows a ring-bus DC microgrid. In a ring-bus network, even if a fault occurs on the network, a second path is available for the power flow.

In commercial premises where there are equipment and appliances operating at different voltages, a DC microgrid with multiple DC buses is used. A common industrial standard for a high-voltage DC bus is 380 V, whereas low-voltage DC buses of 48, 24 and 12 V are used [6].

☐ Questions

1. In the MVDC scheme shown in Figure 11.4, the voltage of the infinite network is 11 kV (L–L) and VSC terminal voltage is $\mathbf{V_c}=15\angle20°\text{kV}$ (L–L). Calculate the active and reactive power transfer to the network. The value of the inductor (L) is 50 mH. Neglect all losses.

2. A PV array is connected to an AC system as shown in Figure 11.19. The inverter maintains the voltage at busbar C at 325 V. The PV array has 16 modules in series and when the voltage across each module is 18 V, each extracts maximum power. The total power output is 4 kW. Calculate the duty ratio of the DC–DC converter that ensures maximum power extraction. Assume that DC–DC converter is a buck-boost converter. Neglect all the losses.

3. A three terminal MVDC network is shown in Figure 11.20. Resistance of all cable is 0.05 pu/km. The voltage at busbar A is maintained by the

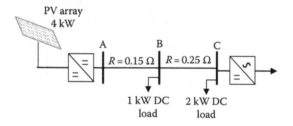

FIGURE 11.19 Network for Question 2.

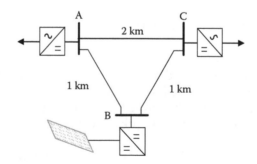

FIGURE 11.20 Network for Question 3.

local inverter at 1 pu. At a given operating condition, the PV plant generates 0.8 pu and the power flow at A in line AC is 0.25 pu from A to C.

a. Calculate the voltages in the other busbars and the flows in the other lines and losses in all lines.

b. The inverter at C operates on a droop control with 1 pu voltage at no power transfer. What is the droop of this controller?

c. If the PV power output of inverter C changes to 1 pu, what would be the new voltages and flows?

☐ References

1. W. Zhang, H. Liang, Z. Bin, W. Li and R. Guo, Review of DC technology in future smart distribution grid, *IEEE PES Innovative Smart Grid Technologies*, Tianjin, China, May 2012, pp. 21–24.
2. D. J. Hammerstrome, AC versus DC distribution systems – Did we get right? *IEEE Power Engineering Society General Meeting*, Tampa, FL, June 2007, pp. 24–28.
3. J. Arrillaga, Y. H. Liu and N. R. Watson, *Flexible Power Transmission: The HVDC Options* [Hardcover]. Wiley-Blackwell, England, 2007.
4. D. Mehrzad, J. Luque, M.C. Cuenca, *Vector Control of PMSG for Grid-Connected Wind Turbine Applications*, Aalborg Universitet, Denmark, 2009.
5. Y. Yang, J. E. Fletcher and J. O'Reilly, Short-circuit and ground fault analyses and location in VSC-based DC network cables, *IEEE Transactions on Industrial Electronics*, 59(10), 3827–3837, October 2012.
6. T. Dragicevic, J. C. Vasquez, J. M. Guerrero and D. Skrlec, Advanced LVDC electrical power architectures and microgrids: A step toward a new generation of power distribution networks, *IEEE Electrification Magazine*, 2(1), 54–65, March 2014.

INDEX

9 780367 573874